W9-CNH-156

Books by Stanley Keleman

LIVING YOUR DYING
THE HUMAN GROUND
SEXUALITY, SELF AND SURVIVAL
TODTMOOS

Stanley Keleman

YOUR BODY SPEAKS ITS MIND

The Bio-Energetic Way to Greater Emotional and Sexual Satisfaction

• • •

Simon and Schuster
New York

PUBLISHED BY SIMON AND SCHUSTER
ROCKEFELLER CENTER, 630 FIFTH AVENUE
NEW YORK, NEW YORK 10020

DESIGNED BY ELIZABETH WOLL
MANUFACTURED IN THE UNITED STATES OF AMERICA

1 2 3 4 5 6 7 8 9 10

LIBRARY OF CONGRESS CATALOGING IN PUBLICATION DATA

KELEMAN, STANLEY
YOUR BODY SPEAKS ITS MIND

1. BIOENERGETIC PSYCHOTHERAPY. 2. BODY, HUMAN.
I. TITLE. [DNLM: 1. CONSCIOUSNESS. 2. SELF CONCEPT.
BF311 K29Y]
RC489.B5K44 616.8′914 75-14407
ISBN 0-671-22095-0

My friends whose books have shaped me:

Karlfried von Dürckheim: *The Japanese Cult of Tranquillity; Hara; Daily Life as Spiritual Exercise*

Nina Bull: *The Attitude Theory of Emotion; The Body and Its Mind*

Alexander Lowen: *The Physical Dynamics of Character Structure*; *Love and Orgasm*; *The Betrayal of the Body*

Ola Raknes: *Wilhelm Reich and Orgonomy*

J. Samuel Bois: *The Art of Awareness*; *Breeds of Men*; *Epistemics*

David Boadella: *Wilhelm Reich: The Evolution of His Work;* ed. *The Journal of Energy and Character*

And I'm deeply indebted to Jack Grant,
who helped me write this book.

For my parents,
Rose and Joe

and for my friend at Todtmoos,
Graf Karlfried von Dürckheim

CONTENTS

OUT OF THE OCEAN

OUT OF THE OCEAN

I come from the bioenergetic tradition, which taught me the importance of the body. I learned that the form and the movement of my bodily expression reveal the nature of my existence. I learned that I am my body. My body is me. I am not *a* body; I am *some*-body.

In doing my work, I have found this to be absolutely true. Graf Dürckheim, with whom I studied for several years, put it this way: "The body you have is the body you live." Our feeling and responsiveness shape our lives. We form our bodily selves as we shape our own reality. Our bodily living shapes our existence.

Your body is not only feelingful, but formative. The continual thrust toward shaping your living insists on your being more—more contactful, more interactive, more satisfied, more of your self. Your formativeness is a cornucopia of urges toward enrichment and fulfillment.

• • •

Sleeping and rising, lying and standing, resting and walking is the primordial pattern of our consciousness. It's a rhythmic pattern. We correspond to the rising and setting sun, to day and night. We match the rhythms of changing days—full moons and crescent moons, spring tides and neap tides—with the rhythms we feel in our own bodies: with feeling excited and feeling tired, with waking and dreaming, with semen and menses, with living and dying.

But standing—what can that be compared to? The first man who stood up must have felt very different. Standing made him different. Being upright organized him differently, formed him differently. It made for a new life-style.

To stand is to walk. Being erect extends our ability to satisfy our hungers—for food, contact, warmth and shelter. Our uprightness begins the process that we recognize as human consciousness, humanness. To be upright is to reduce chance, eliminate randomness. We stand and we increase commitment.

When we're on our feet, we focus and express our selves. Standing up shifts our emphasis from experience to expression. Experience, while present in expression, does not demand focusing or risk. Experience fills and expands us; expression shapes us. Being vertical rather than horizontal changes the stream of our sensations, changes the orientation of our nervous system, makes us more interactive, more alert, more individual.

Being upright changes our relationship to the ground. We are no longer land-fish, swimming along on all fours. Now we walk. It's ironic that to be erect we must grow down. The legs unroll, and the spine lifts the head.

To become somebody starts with being aware. Being aware starts with standing, or the urge to stand. Standing starts with finding one's ground, which shapes our bodies. This occurs by the development of attitudes which help us to organize our living.

Life speaks to us in the words of the living—as feelings and sensations, as hungers and needs. Our bodies are the wellspring of these words. We speak to ourselves in the language of who we are becoming or struggling to be, the language of how we are forming.

From the pool of life I have been formed into a man. From my own forming, I write this book.

BODILY ROOTS OF AWARENESS

I lived an extraordinary event. I took a ship from Hamburg to Los Angeles. I took a ship because I didn't want to fly; I didn't want to be dumped back into the culture in twelve short hours. I'd lived in Germany for about three years, and I thought I'd take a long, leisurely boat trip. What fantasies I had: thirty-one days—nowhere to run; I figured I'd have a ball.

So I boarded the ship, and the steward said to me, "Which cabin do you want?"

I said, "What do you mean, which cabin do I want? Don't I have an assigned cabin?"

He said, "You can have any one you want. You're the only passenger."

Thirty-one days . . .

The captain and I were the only ones taking formal meals. Once in a while the first mate came in. We ate breakfast at eight o'clock. The second morning out, I got down just in time to see the steward turn the clock back fifteen minutes. "We eat in fifteen minutes," he told me.

I said, "But it's eight o'clock."

"Nope," he said, "we just passed the meridian. It's gotta go back."

This went on every day. I'd come down and find the steward turning the clock back. It meant that I ate breakfast later and later, because I was still getting up according to the schedule I'd been keeping when I first got on board. But there came a point when I refused to have breakfast four hours after I got up. I started to eat when I was hungry. I was the only passenger; for five bucks the steward was my friend, and he fed me when I wanted to eat. I went to the table to keep the captain company, but not for breakfast.

So I found myself on my own clock. And the more I began to feed myself when I was hungry and sleep when I was tired, the more I began to experience the world in an entirely different way. I became the me who was the ocean, the me who shaped the semi-permeable membranes separating my ocean into inner and outer, the me who made me me.

When I was satiated, I went to sleep. I woke up hungry, recharged with excitement. "Oh," I said to myself, "this rhythm of waking and sleeping, standing and lying down, has to do with being excited. Excitement focuses and reaches out. Being up, being awake and standing, has to do with being excited; and going to sleep has to do with being unexcited."

During those thirty-one days on shipboard, I began to understand that excitement is rhythmical and self-initiated. I stand and I'm active; making my world, being me, has to do with my being up. And then I lie down in order to recharge myself. The up and down of my daily life is not opposition but complementarity, presenting the two aspects of my excitement that form me, shape me. When I'm asleep, I assimilate the past day. This gives shape to my night—just as the field of the night, with its reveries and dreams, gives shape to my new day.

• • •

We talk a lot about awareness. What is awareness? What is consciousness? For me, our awareness is our living process. I don't believe that there is a conscious part of us which directs our behavior, independently of our bodies. What I believe is that our biological process—expressed as moving, feeling, perceiving, and making patterns of meaning—forms our field of experiencing that we call knowing. Our perception of our own life activity is what we call awareness. What our lives are about is what awareness is.

If I imagine myself to be an oceanic field, a web-patterned sea which oscillates and pulsates, my awareness is that part of me which is excited. Sometimes I know who I am and sometimes I don't recognize who I'm becoming, depending on the intensity and the location of my excitement.

We know ourselves as our excitement comes forth in the forms of action and feeling, expectation and desire. Our greatest energy is at our surface, which we are continually forming and re-forming. Our boundary is the place where our excitation defines us, where our experience emerges as us—our shape, our particular person. The primordial ocean swells up in a wave, forming individuation.

The ecstasy and anguish of the human situation is that we live and perceive the form of our life in the immediate present. And, standing up as we do, we are able to see into the distance, the future. And so we live these two aspects of now at the same time. No other animal's excitatory patterns extend the now in this way. No other animal is so full of the energy that reaches out and shapes that which is not yet there.

Waking, standing, and walking take us out of the ocean, that unbounded space where we are pure expansion. On the ship from Europe to America I discovered that my awareness is actually the process of my excitement thrusting toward form. The unbounded finding boundaries is what awareness is.

The continuum of our self forming is what we are conscious of. Lying down flattens and broadens our excitement, our awareness, our self-feeling, making our boundaries less differentiated. Standing up intensifies our excitement and deepens our awareness, our self-feeling, as we reorganize our selves in the sphere of gravity.

Standing is the thrust of our self forming, of individuating our lives, of creating a life-style. In getting up we shape the space for new response and new focus. We experience new forms of pleasure, new patterns of expressing our love for another. Sound becomes speech and acquires the distinctions of language. The emotions of waking and standing form the human body, the human awareness. The awake body *is* our consciousness.

Our Awake Body

There are two heroic and major phenomena that occur for us every day of our lives. The first is getting up in the morning. And the second is going to sleep at night. Simple events. We live our lives within the framework of these two events.

The human animal has evolved from being on his belly to being on his feet. Our evolution embodies two aspects of living: the living of our former horizontal stability, and the living of our present vertical instability, our unfixedness and responsiveness. This unstable and highly responsive standing is the contemporary expression of the evolutionary drama.

When working with people, I usually ask them to lie down, to relieve the effects of gravity. Then, later on in the session, I ask them to get up from the floor or the bed with their experience, to receive the effects of gravity. Being upright leads to a new feeling, a new standing self.

Through my work I have learned how people are on their backs and bellies and how they are on their feet. On their backs and bellies, they are more helpless, more subject to chance. On their feet, they're more in control, even though it may be more risky. Animal horizontality gives contact with the ground; it is expressed in four-footedness, with the head and torso in the same plane. Animal verticality, while diminishing the area of contact with the earth, intensifies the connection and also opens an expanse of belly and chest for meeting the world.

Man's ancestors—who first engaged in the pulsatory, wave-like movements of swimming, and then in four-footed swimming on land—were secure in their patterns of locomotion. It was when man stood up that his relationship to the earth became insecure. And that state of unsureness and instability serves as the foundation of human consciousness. Human consciousness is part of an energetic process that has a pause, a holding-up of action for the fraction of a second in which we form our next move. Human consciousness is the heightening of energy that comes in this brief pause.

Our vertical extendedness gives rise to forms of feeling and locomotion, patterns which express the human condition. And the human condition is human awareness. The reality of our uprightness is what we are about. It is an ongoing process to which we commit ourselves. We choose to form our uprightness.

The association between verticality and awareness is recognized by many spiritual disciplines. In order to enhance the energy for consciousness, these disciplines prescribe the maintaining of a straight spine. Western psychoanalysts use horizontality to get at the source. They ask a person to lie down so that he can feel his dependency and helplessness. They ask him to return to his vertical stance armed with the energy of new insight—which, they hope, he will translate into new patterns of activity.

• • •

There is a biological process called "ontogeny recapitulates phylogeny." The foetus repeats, on its way to being a human, major stages of evolutionary development. It takes on shapes that are continually changing from conception to birth, as one lives through the history of cell life, fish life, amphibian, mammalian, and human forms.

I would like to suggest that ontogeny also recapitulates phylogeny *outside* the womb, in the drama that takes place between child and environment. During the first three years of this drama, the child learns to go from a horizontal position to a vertical position; and it is probably, along with acquiring the use of verbal language, the most important achievement of its life. Can you imagine the amount of energy at hand for learning how to stand? In order to get itself on two feet, the organism has to rip itself out of the horizontal environment, the dependent environment. If the horizontal environment is poor, the organism tends to remain dependent, depressed, down. It cannot become independent as long as it is lying down.

I equate verticality with individuality. There is a three-way connection between verticality, our capacity for higher excitement over a sustained period of time, and our increased ability to make distinctions and selections. Man is the most highly individual animal on this earth, and the most consciously selective and environment-altering. Our uprightness presents nature's forward-goingness. We pick our own path instead of being driven.

A higher energy metabolism leads to a more lively connection with the world and a wider reality: a greater grasp of what is and of what is possible. We have all seen folks who collapse, who lose their uprightness and have lowered aliveness. Collectively, people with higher energy create new social forms. If people try to live in a recumbent position, they flatten their excitation and deny their emerging individuality.

We who stand are the only animals capable of loving. Other animals have contact, have connection. But the development of richer, more tender relationships as a consistent possibility in one's existence depends upon standing and exposing the tender side of the body. With four-footed animals and animals that crouch over, the front of the head is the leading edge. They receive the world with sight and smell. But for the human being, the whole front of the body is the leading edge—not only the eyes, nose, and ears, but the chest and belly and sex organs. All of this warmth and extended touching is now the leading edge. That's what it means to stand up. To stand up is to open up, open out.

We are not encountering the world with just our eyes and noses; we're encountering it with the entire front of our bodies. The front of us is an extended surface of contact and connection. This is what we present to the world. We might say that the front of our body extends the surface of our brain, or that our brain extends itself as skin and muscles, organs and nerves, which deepen our connections with others and shape fresh satisfactions.

Being Up Front

When we stand, we expose the underside of our bodies. We hold ourselves open to the world. The underside of us, formerly protected, now faces outward. Our softness, our tenderness, is exposed to the environment and to other people. There is a deepening and broadening of contact which says: "I am willing to risk, to meet, to accept. I permit myself to be influenced. When I feel vulnerable and threatened, I contract; I narrow my life space. When I no longer feel threatened, I open up again and take more space."

When we express ourselves, we are not organized in a static relationship with the world. We organize in a fluid way. Static relationships stereotype energy, inhibit excitement. Our two-footedness expands and intensifies our excitement, our responsiveness.

A dynamic, responsive, upright attitude does not permit the stereotyping of responses. It encourages excitatory responsiveness. When we are scared of being excited, we cover up with muscular rigidities that give us the illusion of strength. A person may lock his knees to produce the illusion of stability. When he unlocks his knees, his responsiveness and the fear of his responsiveness emerge together.

The act of waking and standing up makes the world pulsate with excitement. Being upright makes the world asymmetrical. One takes a step; one goes up and one goes down. One has an erection and one does not. Systole and diastole. Our actions increase and decrease charge. They increase and decrease desire. Think of a lover, a runner.

When you are engaged in a sexual encounter with another person, encouraging your feelings, you let the heart open up, you let the unexpected occur. You are expansive; you're moving toward the other, not knowing. You are willing to involve yourself. There is no performance.

Fixed forms, absolutes, are an illusion. Everything

in nature indicates that nothing is permanent. When I work with people I don't attempt to make them into anything. I try to help them experience and be more. What happens when we accept our continual forming instead of seeking permanence? We discover that our lives are an adventure, an emotional odyssey.

Maturity is being willing to enjoy one's own self-perception. Maturity is the willingness to stand, to shape one's self, rather than compulsively leaning on others or on a set of ideals.

In talking about the development of human awareness, I'm not just talking about the mental life. I'm talking about the willingness to accept feelings and sensations. I'm talking about the ability to refuse conditions on love, definitions of love, and to accept the experience of what love is. I'm talking about people discovering the feeling and the experience of their own rhythmicity, and then going on to discover the particular rhythm of their gravitational relationship.

We are always involved with a relationship to gravity, always intimately involved with a relationship to space. A large part of the nervous system is devoted to dealing with this. Gravitational and spatial relationships are the template of social relationships. Gravitational and spatial relationships organize social forms and connections. Being upright generates the *human* connection.

How can I fully transmit the feeling of freedom in human experiencing, which is represented by our standing up? In being erect we are free to look beyond the edge, unbounded by old images and old forms, free to be undulating toward a new expression, free to be strong enough to take that next step, free to breathe, to generate our own awareness rather than introjecting somebody else's knowledge. When we take our time to wake up, we discover a world that is so . . . I don't know what—even if we can't make head or tail of it. And anyway, who the hell are we satisfying? Who are we living for? For whom must we have answers? Who are we making order for? We have all been conditioned so thoroughly to prove ourselves, to make ourselves correct. But we're not in school anymore. Stand up and be your self.

GROUNDING AND BODILINESS

Grounding

Grounding is an expression of our planetary life. Grounding connects our processes of excitation with the earth, forming us both. Just as the groundedness of a tree routes the flow of sap from earth to leaves and from leaves to earth, so does our groundedness channel the flow of excitement from ourselves to the environment and from the environment to ourselves.

This flow of excitation nourishes us and intensifies our connectedness. Grounding establishes the circulation of our streaming sap, the streaming of our blood. It sets up a rhythm of ebb and flow and a vibrating resonance with our environment.

We have all heard expressions such as "having one's feet on the ground," "being well grounded." What is it to be well grounded? How does one find one's ground? A tree, in biochemical relationship with the earth, sends forth tentacles called roots which make it part of its ground and the ground part of the tree. A child, in biochemical and emotional relationship with his parents, finds his ground by growing into it, by learning to stand on the earth and to move with his feet on the ground. Anyone watching a child walk can see whether it's secure or insecure. Just as the tree interacts with the earth, grounding itself through its roots, the child interacts with his parents and sends down his own roots, called legs. The biochemical interaction is also a social and linguistic interaction. If the soil is poor or the weather harsh, grounding may be poor or too sinewy, too tough.

Just as a tree can be uprooted, so can a person be uprooted. Windstorms uproot trees, and emotional storms uproot men and women. Emotional storms break the continuity of excitatory flow between ourselves and our environment—the continuity which grounding provides. Connectedness with our biological ground allows the circulating of vitality, giving rise to love and growth. Separation from our biological ground results in fear, rage, anguish, and even death. A Kalahari bushman dies when taken away from his native land.

Grounding grows out of being born. It grows out of coming into the world with a body. We plant ourselves in the world. Our natural function grows roots on one end and leaves and branches—social relationships—on the other.

• • •

Some of us are so embedded in our family or our tribe that we take our living for granted. We live quite unselfconsciously. When we lose our homes or our traditions, when the storms of tragedy and new growing tear us from our connections, we begin to be aware of what it is to have our ground, to have our legs and be able to move around, to have that place from which our nourishment springs in an uninterrupted flow.

We can be grounded without being conscious of it. There are those of us who work and love and interrelate with the world, yet who never have the experience of being formers of their own world and of their own selves. It is not until our way of grounding ourselves is threatened or significantly altered that we begin to appreciate how we connect with the world. Though emotional storms may uproot us, weaken our bodies and personalities, they may just as well serve to deepen us—to make us more vivid, more intensely us.

Being Bodied

To be born one must have a body. To die one must give up his or her body. Our bodies are us as process, not us as thing. Structure is slowed down process. As life builds structure, it builds itself in. How alive we are, how deeply responsive and expressive we are, shows in the graceful shape of our body, which reflects our connectedness of feeling, thought, and action. How unalive we are, how unincarnated we are, reveals itself as unresponsiveness, ungracefulness, and restricted bodily mobility.

To be grounded is to establish a relationship with the earth. To be embodied is to establish a living body—not just to be *with* your body, or in relationship to it. Your living body creates your relationships.

How we live our bodies is the story of our process. Our excitement tends to create boundaries or a capsule to embody itself. Our stream of excitation inhibits itself at crucial points in its cycle of development. The thrust of our excitement triggers a self-inhibiting which retains itself, so that our excitement does not fully lose its shape. It collects itself, contains itself. It forms a boundary, a capsule, an image, a body. This is the development of the organization that we perceive as "us."

I have a film* that shows very clearly how protoplasm is capable of forming structure from itself. The protoplasm is pulsating, streaming. One layer of the streamings thickens and creates a membrane which acts as a channel for the main flow, giving it more form. This containing creates an individuation of velocity and rhythm. The different rates of excitement and the asymmetry of vibrating and resonating qualities result in a body. The protoplasm has been bodied.

Our different excitatory levels generate our various experiences in living, the joys and sorrows that form our personality. Being bodied is the shaping of our living flesh, the forming of our self as a living somebody.

• • •

We can interfere with embodiment by not permitting boundaries to form—or by not permitting boundaries to unform. Either way, we can discourage our future, our self-forming.

There is a disease called hospitalism. When a child

* "Seifriz on Protoplasm"

is born, if it doesn't have a mother and if there are no mother substitutes, it is put in the hospital and virtually ignored. Ignored, the child very often becomes apathetic and dies. Its streamings of excitation experience no contact, no responsiveness. The organism senses that it has no ground—no mother to ground itself with. And without a ground it has no future, so it terminates its own processes and collapses its own boundaries. It prevents its own structure from unfolding.

The forming of boundaries necessitates a pre-self-conscious, pre-personal decision. This pre-personal decision lays the groundwork for later boundary-formations that are individualistic and personal. Some people, because of very negative conditions in early life, were able to form only partially. During their early years it was too painful for them to fully inhabit their flesh; and so they decided not to body themselves fully. They formed themselves in a diminished way, and although they may now be adults, we recognize them as being babies. Schizophrenics live another kind of diminished human existence: part human, part shadow; part social, part unsocial.

Those of us who do not inhabit our flesh, who do not have the deep satisfactions that our bodies can give, are always banging at the door of ourselves, trying to get satisfaction. Those of us who are afraid of our impulses lock ourselves in a world of ideas.

Those of us who are continually bounding and unbounding, forming and unfolding ourselves, feel neither trapped nor lost. When we don't confuse ourselves with a social image we form a bodily self, a somebody, from our pleasures and satisfactions, our pains and sorrows.

I Say No

To say *no* is to make a statement of protest and self-affirmation that heightens one's excitatory processes and vivifies the sense of "I." In the early life it occurs spontaneously, circumstantially, pre-personally. The child's body stiffens or holds a certain posture, and in this pattern of expression its character is formed. I recall how my infant daughter began her expression of *no* by turning crying into screaming, then stiffening all over so as not to be movable; thus her stubbornness was born. A few months later she shook her head, set her jaw, and said *no*.

If I wish to maintain my individuality, my life space, I must accept the pain and the pleasure that come with risking and taking a distance from my support. *No* communicates my willingness to risk distance, separation, and loneliness.

Many of us have a problem saying *no*. Even more of us have a problem saying *no* and making it stick. Or else we say *no* so rigidly that we are then incapable of allowing the *yes* to occur—the pulsatory movement that reconnects.

If you do not say *no*, you never affirm yourself. If you don't exercise the ability to form and maintain boundaries, you become victimized. Of course, if you negate the world, reject others so that only you exist, you will lose your self too. But you'll never be yourself unless you're willing to cut your self off from the source—which may be mother, which may be the culture, which may be the peer group. You may even have to form a contraction. A woman I worked with told me that, since she could not fight her father or run from him, she had to stiffen up all over in order to keep him from intruding on her space.

In many cases, a contraction is the strongest expression of self-affirmation that a child can make. A child says *no* in order to protect and assert itself. And if that is not respected, do you know what happens? You get a very amiable bowl of jelly, a *no-body* that cannot bear the excitatory processes that form independence.

The contraction, the *no* which inhibits expansion, at the same time affirms. But it may become so firmly embedded that the person is taut, won't open himself. Then he may come to me and say, "Help me to again be responsive. Help me to trust again, to uncontract, to unbound, to learn to say *no* in a different way." To say *no* first distances you, and then permits the expression—the *yes*—of new action, new you.

A contraction does not have to be a chronic muscular cramp. It can be a temporary set of personal decisions. The formative process requires that you set boundaries and form yourself—then soften your boundaries and re-form yourself.

VIBRATION, PULSATION, AND STREAMING

The Stuff of Creation

I asked a woman I was working with to stand up and breathe in such a way as to prolong her exhalation without tightening her abdomen. After a bit, she felt afire; she was shaking and tingling. These vibrations, sensations, developed into rhythmical contractions which then, to my eye, became a pulsatory series of electric explosions. She expressed these waves of feeling as tender movements and sounds. I felt in me a responding vibration of tenderness, and then, as her expressions increased in assertiveness, I experienced the rhythmical intensifying of my tenderness and a softening, which became a stream of softness reaching across space, connecting this woman and me in a river of excitement and feeling.

The way in which we perceive the world and interact with it depends fundamentally on the quality of aliveness of our tissue. Our tissue tone—its health or its ill-health, its vibrancy or its deadenedness—is the background of our experience and our perception. We all know how a healthy baby feels to our touch. And we all know how a sick person feels. We associate a hard tone with a he-man, and a flabby tone with a weakling.

There are three states of aliveness: vibration, pulsation, and streaming. Each state has qualities that are distinctive and directly observable, although one shades into the next. Our bodies display all three of these states. Vibration, pulsation, and streaming are natural functions of protoplasm, of cells and organs—natural functions which can be seen under a microscope. They may also be experienced subjectively as qualities of feeling.

I feel the universe as a continuum of vibration, a shimmering field of excitation. This vibrating field gives rise to an increase in excitation, which tends toward expansion; and the expanding excitation triggers a self-inhibiting mechanism which limits it and forms its boundaries. The excitation continues to swell against these boundaries until it cannot swell any further; and now there is a slight shrinking, a beginning of self-collecting, a sort of jelling or clotting. This is how the quality of pulsation develops. Actually, the initial expansion is already expressive of a pulsatory state, but the pulsating is not seen as such until the expansion has been bounded.

If you have ever been seriously cut, first you're shaken and vibrating, and then you begin to throb; your world rushes in and out. When pulsations occur rapidly and in series, you have streaming. Streaming is a continuity of pulsation, a stream of rhythmic excitation that maintains itself in a particular direction and in a particularly organized form.

If you hold your breath and pay attention to your chest and abdomen, you will feel the coming and going of excitement. If you clench your fist or tighten your thigh muscles and sustain the contraction, you will feel a fine vibration throughout your entire organism. If the vibration deepens, you'll begin to experience it as a pulsating. Sustain the contraction until the pulsating deepens, then let it go, and you become aware of a streaming: an internal flowing which is difficult to see but which you can feel. Streaming is like the flow of sap in a tree. It feels similar to the rhythmic currents of blood, the rhythmic currents of thought, enhanced by a subjective sense of sweetness and glowing, and—for me, at least—a quality of moving outward into a different space, a different time, a quality of connectedness and knowing.

Two of us are excited. The two of us together intensify our fields of excitement. We begin to expand, to make movements toward each other, gestures back and forth. This is pulsation. Then the feelings of excitement begin to take on a stream of continuity that we experience like an electric current.

Suppose we are at a dance hall. We see individuals with their own auras of aliveness: the dance hall is a sea of excitement. The music starts playing and the aliveness increases. The people make gestures toward one another; they begin to dance. And these dancing movements begin to interact in such a way that, looking on from the outside, we feel waves of excitement sweep the dance floor. We feel the recurrent pulsations of the dancing. The individual dancers have become a streaming organism before our eyes.

If you watch a cell divide, first you see that the cell

is excited, vibrating. You see the forming of two poles, two areas of intense internal activity. You actually see the radiation between one pole and the other, and the lining-up of chromosome bodies within that field. The radiating between the two poles intensifies until it becomes a pulsation and then a streaming. The streaming communicates the deepest information about life—as we do when we communicate with each other. We are all, for better or worse, attuned to patterns of excitement.

• • •

Vibrations, pulsations, and streamings are basic to all human relationships, and to all concepts of freedom and social concern. The child is connected with the mother through these life phenomena. As he develops his own boundaries and his own pulsating, he begins to expand and extend himself away from the mother. He extends himself and reconnects, extends himself and reforms his relationships and his self. In this way, bit by bit, he acquires his personality; he gains his independence. If the child's own streamings are allowed to develop and intensify, he becomes a living example of the paradox of individuality and connectedness.

In our particular culture, however, the normal process of self-separating is artificially speeded up. Our initiatory rite begins at the moment of birth when, typically, we take the child away from the mother and put it in a sterile environment. The rite continues with early weaning and the taboo on sucking, and culminates in the drama of toilet training. These three separative efforts are initiated much sooner in our culture than in others. They help to create a continuum of self-imagery which denies the discontinuous, pulsating life of the body. They lead to acceptance of an artificial schedule, a socially imposed rhythm that kills individual rhythmicity. Wake up at 8. Brush your teeth at 8:05. Eat your breakfast at 8:10 and out the door. Catch the bus at 8:20. Go to school 9 to 5.

The initiation is mainly non-verbal. The attitude of "Don't touch" is communicated directly by pushing away or by holding the child against a hard body. Either act breaks the excitatory connection and thereby hinders the child from developing his own pulsations. By the time the child understands words, by the time he understands the causes of his fear and anguish, the separation has already been accomplished. Then the culture's ideas about the nature of the body and the nature of life find a prepared field to grow in.

• • •

Feelings do not emerge from a background of nothing. They result from movement, containment, and the intensifying of excitation. A person feels the progression from vibrating to pulsating to streaming as a deepening of self-enjoyment and a deepening participation with the rest of the natural world. The streamings brew feelings of rightness and oneness with nature.

Have you ever walked alone into a forest in which there was absolute silence, and in that silence there was so much going on that it almost overwhelmed you? The contact with the streaming is so intense that it quiets the mind, so exquisite that sometimes it becomes unbearable.

I spent my early life in New York City and then

went to Germany to live in the Black Forest. The Black Forest is a thousand meters up—nearly four thousand feet above sea level. I was crazy for weeks. I couldn't take a satisfactory breath, and I couldn't imagine what was going on. Then I made a discovery. My breathing difficulty didn't have to do with the rarefied air or with the number of my red blood cells. It was rather that I had been pulled out of a pollution chamber and put into a clean backwoods environment whose vibrations I was totally unused to deciphering.

It took me a long time to accept the feelings that those vibrations created in me. When I did accept them, I recognized that these feelings were akin to the unconditioned existence that I felt as a very small child. I recognized that they came from the same excitatory sea that I lived in during the pre-verbal part of my life, the part of my life when I was directly experiencing the world in qualities of vibration, pulsation, and streaming.

The world that most of us no longer recognize is the world of vibratory connectedness. When our vibration emerges in us, we read it back as danger—or we treat it like a stranger. We accept this connectedness in certain stepped-down forms: "I need you; I want you; I appreciate you; I approve of you; you make me feel good . . ." But when we are presented with intensified vibration, most of us are unable to accept it. We think we are sick, bizarre. We do not recognize our life at this level.

We Are More When We're Vibrant

Our individuality is our personal rhythm of pulsating. Our relationship to gravity, our back-and-forth dialogue with ourselves and with others, our breathing patterns, our action patterns, our dreams and loves, the qualities of our tissues and organs are statements of our pulsatory individuality. They determine how we perceive ourselves and our world, how we create our values, our needs, and our choices. A person with tissue that is not very motile and a structure that is not very vibrant will feel the world as stronger than he, and he'll either be threatened by it or be glad that it is protective. A person who demonstrates a great deal of vibrancy will challenge the world, or feel in harmony with it—but he won't feel submissive.

We deceive ourselves to think that if we pulsate and vary our self-expression, we are unstable and unreliable and don't know who we are. On the basis of this self-deception, we go on to seek our identity according to the definitions of socially approved roles. We deny the changing patterns of our individuation by trying to maintain an unchanging image. But a rigid identity is not individuality. To affirm our individuality, we have to give up our search for static roles and attitudes and, instead, seek connectedness with our own pulsatory rhythms. To be an individual is to impress the world with one's varying expression rather than merely to mimic the expression of somebody else.

When we identify with our streamings, we discover our own continuity. Pulsations and streamings are discontinuous, yet there's a continuity to them. It's like waves breaking on the shore. The waves are discontinuous, but the process is continuous. There is no continuity that does not incorporate some kind of discontinuity. Living this pulsatory discontinuity destroys stereotypes, demands that we give up the old and create new spaces, new forms, new connections. To deny this discontinuity is an attempt to establish security, permanent possessions, a rigid social structure.

Vibrations, pulsations, and streamings express the sense of excitement and unknowing that develops out of moving from the horizontal to the vertical position. If I am lying down, my pulsating is fairly quiet and stable. I feel the steadiness of a single contact with the earth. If I stand up, my pulsating grows more intense, and my contact with the earth becomes unstable and two-pointed. When I stand I sway. I shift from foot to foot. I move toward and I withdraw; I reach and back away. I know and I don't know. I say *yes* and I say *no*.

Everything in the human being points to discontinuity. We are pulsating all the time. Peristaltic waves flow through the alimentary and vascular systems. The nerve fibers pulsate. So do the biological clocks that regulate the flow of glandular fluids. Laughing and crying, orgasm and ejaculation are pulsatory, rhythmical. The muscles extend and flex. I open and close. I love and I don't love. My feelings come and go. I *am* my discontinuity—my connectedness and disconnectedness.

When two people connect sexually, the deepening of their connectedness has a rhythmic pattern to it. To "push" one's sexuality expresses the need to control one's excitement, to restrict it or to keep from losing it. Pushing results when the socialized performance images interfere with the natural rhythms of bodily connection. These performance images are internalized as chronic muscular contractions which disrupt the excitatory processes. The strength of one's pulsatory waves is diminished. One's responsiveness is diminished.

If my streamings are interfered with, if my freedom is interfered with, I react violently. I do violence to myself, to you, or to my environment. I twist myself. Or I deaden myself, live in a sort of sleep. I curtail my social contacts. When my avenues for connectedness are narrowed, I cover up the pain of the narrowing by giving myself reasons for holding back. I settle for compensatory satisfactions rather than bodily satisfactions. I fulfill somebody else's ideals; I seek society's goals and try to feel satisfied that way.

The Field

One day as I was sitting on the shore of the Zurichsee, looking south toward the Swiss Alps, I saw off in the distance a flock of geese flying across the lake. The cable railway was moving over the bridge in the background. And all of a sudden I felt, vividly, my connection with those flying birds resonate out from inside myself and across the lake. I felt myself resonating with the lake and with the bridge and with the mountains. I saw myself as a point in a huge pattern, a huge field of interconnected form and forming. And then I felt my *self* as the pattern. I felt my self as *part* of the field and *as* the field—able to perceive my surround and being my surround—with everything connected by this resonating pattern of excitement that I was experiencing. Birds and me and water and mountains and me and birds.

There were no secrets. I experienced that the pattern of living is to be forming new patterns. The birds were still birds and yet they were no longer birds: they flew along in a magical rhythmic forming of new shapes, new spaces. The lake, the ripples on the lake, the bridge in the background, the breathing in me, the pulsating of me with the flying geese—everything was forming from moment to moment in this exquisitely vibrating pattern, this exquisite stillness. I recognized everything, and yet everything was unrecognizable. That which seemed to be steady, like the bridge, I experienced as simply having less vibrancy than that in which it was embedded. Experiencing my self and my world in this fashion—as vibrating and pulsating rather than conforming to concepts and images—brought me into a pulsatory, streaming universe.

• • •

How does one become aware of his own streaming? If you have ever run a long distance, or been deeply involved in a crisis or a rock-band concert, you have probably felt the vibrations turn into an electric flow. When you're in love, it happens. There you are, standing in front of your girlfriend, throbbing and vibrating inside —all kinds of electricity running through your bones, flashes of sweet light streaming through you. You can hardly contain yourself. And yet the more you contain yourself, the more feeling you have; and there are streams of connection, of excitement, between you and your love.

One way I feel my streamings is to close my eyes and hold my breath. By doing this I usually feel excitement in my chest and abdomen. When I exhale, I feel the excitement spread throughout me. When I continue my breathing, it follows my excitatory stream, rhythmic electric currents. I lose my images and thoughts—my brain, my body, is a glow of pleasure; everything shimmers.

I've found that to live with my own streamings is

pleasurable. To participate with my own pulsating is to form my own life. After all, connection and disconnection is a fact of existence, but the *willingness* to disconnect is an act of faith. It's an act of faith that I will wake again after sleep, that the person I love will return, that if I hold my breath I will breathe again, that I will have an erection again, that I will always be somebody.

THE FORMATIVE PROCESS

How We Become Who We Are

In the picture *2001*, a chimp finds a bone that he learns to grasp and lift above his head in such a way that, step by step, he is able to accumulate power. He accumulates this power by refusing to let the bone strike the animal until he fully extends himself and can no longer contain himself. Try it: lift a tennis racket over your head, stretch back, and hold it there until the vibration becomes so intense that you cannot contain yourself. Then hit the bed. Your self-inhibition has organized both a sense of power and the power itself. To hold back too long is to freeze. Not to hold back long enough is to dissipate the power and the feeling of power. Appropriate self-restraint is the essence of one's sense of self and one's sense of power. Kubrick conveyed his awareness of this phenomenon for man the warrior. It is also the experience of man the lover.

The miracle and the mystery of my living is that I

organize myself and shape myself. I call this my *formative process.*

I experience the universe as a field of excitation, a continuum of excitement, an ocean of excitatory currents. My excitement is my basic experience of my bodily living.

My excitement swells and expands. This expansiveness has a quality of thrust that gives me the subjective feeling of growing. When I am expanding and growing I am highly charged. If my expanding excitement continues unbounded, my charge dissipates.

So that my excitement will not dissipate, I have a self-regulating, self-limiting function which protects me against total discharge. Fundamental to the human condition is an autonomous self-inhibiting that never does permit total excitatory discharge—complete bodily unboundedness—until I am ready to die.

To repeat: at a critical point in my expansion, there is the triggering of a self-limiting which begins to inhibit my excitement, to collect me, compress me. When my heart is filled to the limit with blood, it automatically says, "Enough!" When I am filled with the richness of my life, I reach a place where something in me says, "No more." And then I begin to gather myself, to digest my experience.

Here begins my boundary formation, my embodiment—the forming of my loop, my capsule, my container. I begin to bound myself. I begin to experience my self discretely, individually. I begin to feel my power. I begin to feel, awarely, the form of my self.

In developing the form of a loop or capsule, my excitement does not stop expanding; it intensifies within me. My feeling of growing gets intensified by being contained. The outcome of this intensification is more self-feeling and self-perceiving, which compounds into self-knowing.

There then comes a critical point at which I let go of my boundaries so that I can express my excitement. My containing capsule, in addition to intensifying my feelings and perceptions, serves as a channel for my self-expression. When I express my excitement, I interact with the world—in new experiences that will once again provoke me to expand, contain, and express my formative self.

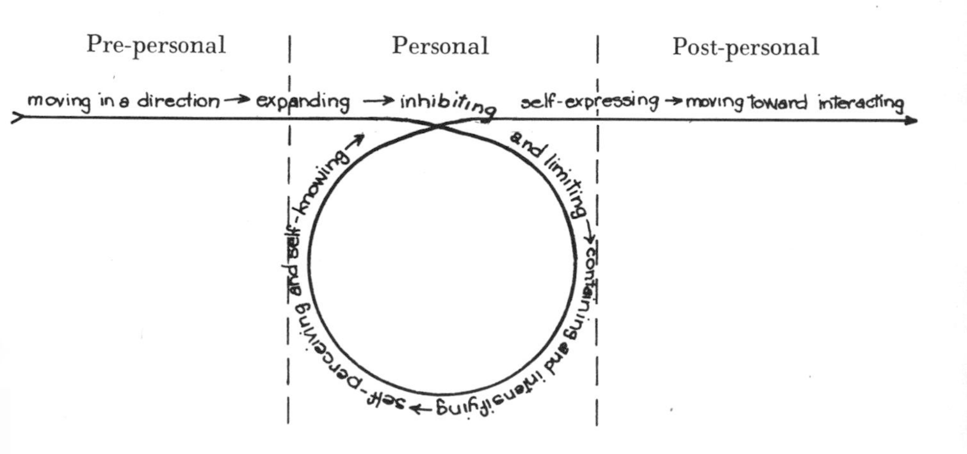

Turning Points

As I live my life, major events form me into being another some-body. These high points of my formative process I call "turning points." Each turning point, each turning of the formative loop encompasses three phases: pre-personal, personal, and post-personal. In the pre-personal phase, my excitement is undifferentiated. In the personal phase, my excitation becomes contained; my individuality and personality emerge. In the post-personal phase, I release my boundaries by expressing myself; and in so doing, I create a self, a field of reality.

When a child is born, it leaves a pre-personal world and develops into a person. When a person leaves the world of adolescence for the world of adulthood, his adolescence fades into impersonality and his adulthood becomes personal. When he dies, he leaves his personal world and lives in the memories of others; his world is post-personal.

Think of situations in your life in which you underwent a major change. Some of these situations originated outside, like going to school, and some originated inside, like the descent of the testes. Think of how these events generated feelings, images, new ways of doing things, new relationships with yourself and with others. Think of how you were before the event and after the event. Think of how the event re-formed you.

When a girl experiences the onset of menses, it's not yet her; it's not yet personal. After a while, the girl expresses her period individually. It's her. When her menses cease at menopause, the experience is no longer hers; it is post-personal.

A turning point has three phases corresponding to the three phases of the formative process. Each transition from one phase to another requires a decision that I make organismically—sometimes consciously and sometimes not.

The first of these decisions is the decision to restrict myself, to define part of myself, to slow my expansion. If I lose myself in my feelings of expanding and do not move toward containing these feelings, then I never end the pre-personal phase of non-differentiated excitation. So I decide to be incarnated, to be born, to be formed.

My second decision is to continue my forming by creating boundaries. I differentiate and become personal. I become human by embodying myself, by shaping my body. If I don't form boundaries, I find myself in a limbo of not being a person, of not having a shape.

I can also maintain my boundaries too long. The intensification of my feeling reaches a level at which I am moved to let go of my boundaries, to give up my present form. My third decision is to become boundaryless, boundless. I decide to leave the personal phase—of adolescence, for example—and I begin to enter the world of adulthood, which is impersonal, post-personal relative to the adolescent world.

At the interface between the personal phase and the post-personal phase, where I am on the verge of giving up my old boundaries, I have the most form. And in that transition from the high point of form to the least form, I compound and intensify my excitation. Think of a man about to strike a blow, or a finger about to hit a piano key. At the point of impact the excitement shapes itself in its most finished form. There is the highest point of form, shape—at impact. Form is released to again form itself. The muscle relaxes and contracts again to make another impact. The greater my excitement, the greater my potential for shaping my own reality, my own truth, my self.

I am not alone in my self-expressing. I share it with

others. My maximum excitement may be the wonderful chord that thrills the audience, or the knockout. My expression impresses itself on others, and the expressions of others impress themselves upon me. It is this interchange of out-pressing and in-pressing which initiates the forming of reality.

• • •

All living seems to be able to form itself. This self-forming is unpredictable and predictable, involuntary and voluntary, impersonal and personal. We all form bodies, and yet each of us forms a unique body.

In the process of forming my uniqueness, I may also form anxiety, because of the risk of my not being able to form again. Anxiety is my feeling when the ongoingness of my formative process is threatened or interfered with. And yet, except when I die, I never totally lose my boundaries. We all give up our form, and yet each of us gives it up in a unique way. We all die, and yet each of us forms his own dying.

Our formative process relates to our dying the same as it relates to our living. One may die in any of the formative phases. There are styles of dying in which people overcontain themselves, squeeze themselves to death. And there are also styles of dying in which people are too willing to unbound themselves, too ready to disembody themselves and move into a fantasy. What this says to me is that our dying is a turning point, an organismic decision that constitutes an integral part of our formative process.

CREATIVE VOID

PRE-PERSONAL PHASE (undifferentiated, expanding excitation)

PERSONAL PHASE (containment, embodiment)

POST-PERSONAL PHASE (expression, interaction, new reality)

CREATIVE VOID

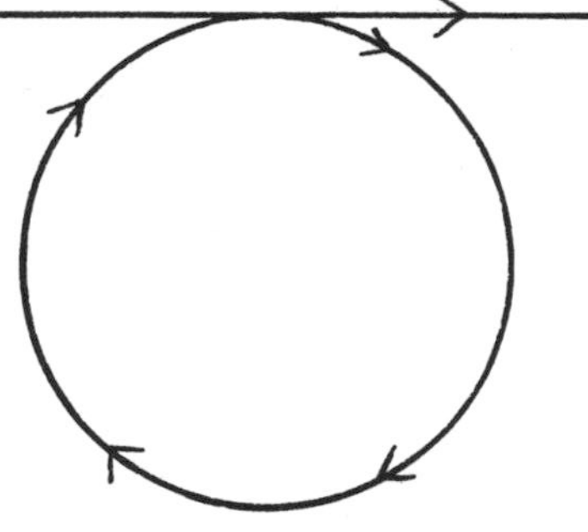

Decision #1 occurs as the line bends into the loop.

Decision #2 is the maintaining of the boundary.

Decision #3 occurs as the boundary is given up.

The overly excited person never puts anything together. He lives in his imagination.

The overly contained person lives imprisoned by his form.

The overly expressive person is always performing. He lives in his actions.

The most intense, most highly charged moment of the formative sequence is the moment of expression. For this reason, the sequence may be re-drawn as a process of ascending toward expression. One needs to keep in mind, however, that the ascending and descending sides are not separate, but connected as intimately as the two sides of a Möbius strip.

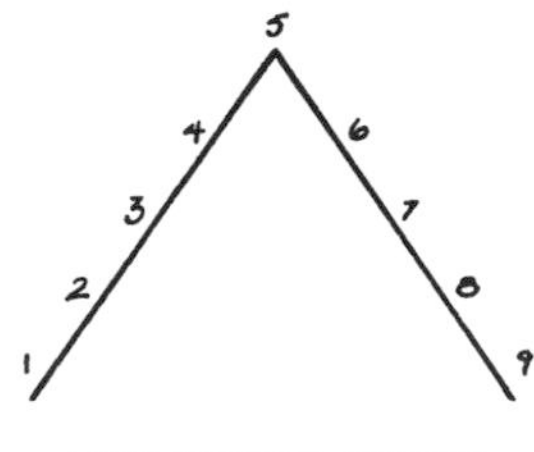

CREATIVE VOID

(5) expression and extension; new interaction; contactful acceptance of the unknown

(4) increasing focus upon goals or avenues for expression

(6) less focus, less precise vision

(3) selection and rejection; choice of attitudes, skills, values

(7) allowance for multiple possibilities

(2) incipient re-orientation; exploration and information gathering

(8) letting go of form; accepting the unknown

(1) insight; new sense of direction

(9) re-forming the world

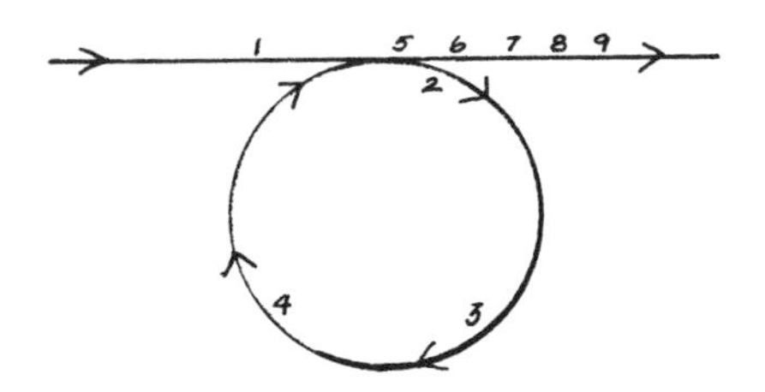

One may encounter obstacles or entrap one's self at each stage of the ascent or descent. When this happens, the process becomes one of ascending and descending frustration rather than an ongoing, pulsatory movement toward fulfillment.

ATTITUDES AND THE FORMATIVE PROCESS

Our formative self is that unique set of attitudes which has never existed before. This human response is the act of creation.

I was walking through the streets of Basel, in Switzerland. And I remember so very clearly saying to myself: "Let your shoulders down, Stanley." Nothing happened. Then I said, "All right, let your *self* down." As soon as I said it that way, I was clothed in waves of excitation. The shift in language made me recognize that my shoulders were holding me in an attitude of fear—I was holding myself with my shoulders. And when I released my shoulders and let myself come down, I was flooded with a tide of excitement, a great ecstatic global feeling.

At that moment the world became vivid, and I realized I was connected with everything. Yet my social identity, all the images and thinkings that I considered as my inner continuity, did not evaporate. I simply discovered: "Hey! I'm more than what I thought, more than what I've been assuming, more than what I used to feel. I'm not merely my opinion of me."

I found myself in a sea of connectedness with all the things and all the people around me. And just as an airplane flies up into the invisible continuum of air and steers with its rudder and maneuvers with its wing-flaps, I felt that I could navigate wherever I pleased in this ocean of connectedness.

I felt myself expansively immersed in the world. And I could see that everybody else was in the same ocean I was in, except that most of them didn't know it. They didn't know it because their muscular tightnesses and their held-in breath tended to diminish their space. They weren't permitting themselves freedom of expansion and expression. In this ocean-continuum, they were trying to maintain the identity of their living space by means of cramped attitudes.

• • •

We generally think of an attitude as a mental set. An attitude is a bodily set. Our attitudes are the framework of our form.

Attitudinal patterns are unlimited in number, and their interaction is simultaneous and complex. Attitudes have muscular, emotional, and mental components. The patterns of our excitement manifest themselves as action, as feeling, and as thinking.

Attitudes form the background for character. In the playing of a football game, the players and the plays, the formations and the styles are like attitudes. They set the limits for how the game is played. The quality of play—the expression that emerges out of the playing of the game—is its character. One side is recognized by its consistent character of being a team of rushers; the other fades under pressure.

A limited number of plays, no matter how well executed, makes for a limited game. If I have rigid attitudes, they not only define my rigid belief system; they also define the rigid feeling system and the rigid acting system appurtenant to my rigid bodily set. It is not just my thinking that is densely bounded. It's my whole body that cannot move freely, that cannot feel freely.

My attitudes combine to form boundaries which contain and express my excitement. If my ability for self-expression is severely limited, it is because I have developed attitudes that restrict the expansion of my excitation. I clench muscles in my hands, arms, mouth, chest, belly, legs. This makes me cautious, conservative, mistrustful. I look for traditions to cling to. I believe that holding on to the known is safer and better than doing things in my own way.

When we are excited and our excitement is accepted and supported, we develop attitudes that extend our boundaries. We reach into the world. We expand the arms, the torso. The heart opens. We feel expansive, and believe that the world is friendly, inclusive of us.

Fulfillment and Frustration

We shape ourselves by organizing attitudes—patterns of preparing for doing and of actual doing. There are two distinct kinds of attitude. One kind is fulfillment-oriented; the other is frustration-oriented. Both fulfillment- and frustration-oriented attitudes attempt to serve instinctual and social needs and manifest themselves in feeling, thinking, and action.

Self-forming organizes my attitudes. If I am fulfillment-oriented, the organization of my formative process grows out of my natural needs. Attitudes of fulfillment are characterized by a forward-going, upright, balanced, flexible bodily shape. There is a symmetry which is revealed as harmony between the left and right halves of the brain—the practical side and the intuitive side—and, more generally, between the left and right sides of the body. The eyes are coordinated. The legs, arms, and torso are integrated. The thoughts and feelings are consistent with the actions of the person, just as there is a form, a grace, a quality that we recognize as an individual, a somebody.

This harmony, integration, and connectedness is perceived mentally as interest, self-confidence, imaginativeness, and a willingness to live with the unknown. It is perceived emotionally as feelings of excitement, anticipation, love, and joy. Think of the visionary. Think of people who are cooperatively committed to a task, people who love to do what they are doing, who feel of service to themselves and to others.

Interference with our self-organizing results in breakdown of fulfillment-oriented attitudes. We become asymmetrical and there is loss of grace and connectedness. If both halves of the body are not involved together in the same action, then we are doing two actions at once: for example, grasping and pushing away, or holding on and trying to move ahead. Caught in a frustrational pattern, we register thoughts of unsureness, incompetence, and confusion, and feelings of resentment, hostility, worthlessness, and despair. There is a general shrinking of our self.

The increasing dominance of frustrational attitudes

leads to descending levels of organization that indicate less and less flexibility to go on growing, less and less of a bodily ability to go on re-forming the self that one is. Continuing to feel interfered with, we retreat into attitudes that we have already tested—old patterns from childhood such as sullenness and clinging which can become more deeply ingrained with each rationalized repetition. We become repetitive—bored and boring.

The Descent into Helplessness

In the process of self-forming, people invariably encounter obstacles. There is unresponsiveness: a baby reaches for mother and she isn't present. There are impediments: a child puts out his hand for something and his hand gets slapped. And there are self-limitations: I can't spring forward well enough to jump twenty feet; I get a flood of insight, but I have difficulty translating it into written words.

An obstacle poses a threat to me when I experience that it interferes with my patterns of fulfillment. I collide with many unexpected situations in the course of a day, but my experience of some of them is such that it stops me in my tracks. I'm startled. I pause before committing myself to do anything. This attitude of surprise,* which ranges in intensity from hesitation to astonish-

* "When this happens to an organism, that layer in the personality that corresponds to the time period of the trauma goes into a state of severe contraction. He may continue to grow above that layer but the growth is not grounded in what went before the contraction."—Personal communication from Alexander Lowen, M.D.

ment, is my initial response to what I experience as interference. A child who falls down hard always braces himself before screaming. A child who has just discovered something new stands there unmoving, totally absorbed in what is in front of him.

The pause of surprise is fundamental to learning. It is the fixed bodily attitude which opens one to information, to being in-formed. And it can be exploited. Brainwashing and hypnosis both attempt to catch people in this pause, so that suggestions may be laid in. The current style of education begins by being disapproving of movement; the stilling of the children's bodies induces individual patterns of alertness which enhance and sustain one another. The energy of each child's alertness can then be directed into forming the role of the ideal student.

The surprised, vigilant attitude develops into either curiosity or annoyance. Curiosity is the first step toward reasserting one's forward-goingness, for it develops in turn into active investigation, fascination, rapt enjoyment, and integration. Annoyance, on the other hand, is the first step of the frustrational descent, a sequence of diminishing excitement which leads progressively to angry dislike, lonely crying, and the frozen terror of feeling utterly helpless.

Let's assume that a three-year-old kid wants some attention from his mother. He sees her halfway down the street and runs toward her. Suddenly a big dog comes out of the bushes, right in his path. Surprised, the kid hesitates; the dog may or may not be friendly. If the kid decides that the dog is friendly, he might give it a curious pat. He might even choose to stay and play with the dog and find his mother later. Then again, if he

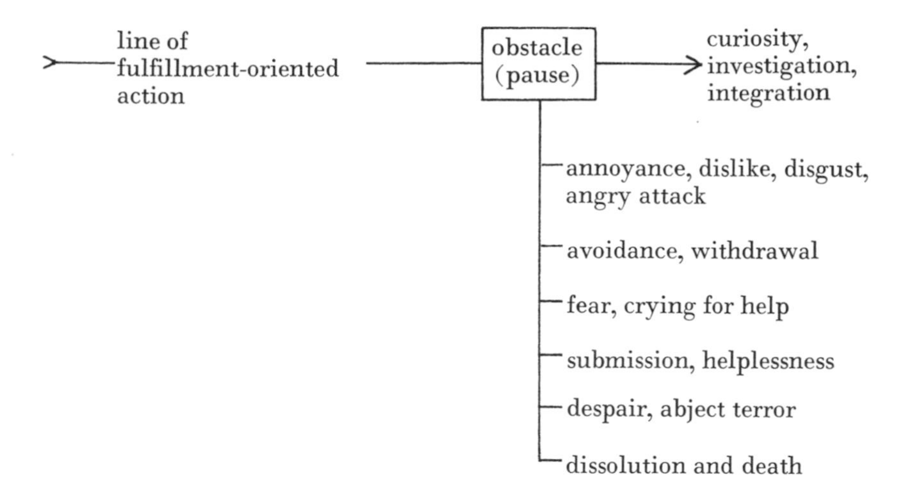

decides that the dog is unfriendly, he might dodge the animal and race to the safety of his mother's skirts.

It may turn out, however, that the kid is caught in the pause of his own surprise. Part of him moves toward his mother, and an equal part of him recoils from the dog. He finds himself stuck in the impasse of trying to divide his energies, trying to run in two opposite directions at the same time. Should he be unable to make a move, he yells for help. If help doesn't arrive, he collapses, surrendering to whatever may come.

• • •

The attitude of curiosity is a vital sign in any society, since it indicates the meeting and resolving of the unexpected. Yet in some cases, annoyance is socially acceptable and curiosity is not. Annoyance is registered emotionally as disgust and mentally as dislike. A young child learns to respond to its feces and its sexual impulses with disgust. A child in school learns to ridicule any appearance of the daydreamer, any evidence of strictly individual perception—in others and also in himself. In each instance, the expressing of permissible dislike is given form by the attitude of impermissible dislike which the child already feels: toward his parents for insisting that he be "clean," and toward school authorities for demanding that he sit still and listen up sharp.

Annoyance is always two-pronged, self-directed as well as other-directed. Suppose I eat something that disagrees with me. I can spit it up, I can run away to bed, or I can do both. But the difficulty is that I still must eat —unless I choose to turn off eating and die. So whether I recognize it or not, my attitude of annoyance expresses a fundamental self-conflict. And if I don't resolve this conflict, I am left with a readiness to reject, often without knowing the reason why. In this way I may gradually lose my creative thrust and begin to perceive fulfillment merely in terms of avoiding interference, eliminating obstacles.

To extend the example of the three-year-old and the dog: the kid may be so frightened that even after he finally gets to his mother, he still feels consumed by the need to be rid of that dog. The attitude of fear has been so deeply implanted in him that for weeks afterward he remains reluctant to leave the house. During this time the object of his fear may become unconscious. But even if the kid knows why he's scared, he goes on feeling helpless to the extent that he hasn't undone the emotional and the neuromuscular components of his attitude. And thus it is that many people live out their lives surmounting obstacles such as resentment or inferiority or poverty—obstacles that have not yet ceased to interfere with their feelings or their bodily shape.

• • •

If an attitude of annoyance does not dispose of the interfering obstacle that has set it off, the attitude continues to be operative. It ensures a closed system. As a closed system of behavior, a mild frustrational attitude can easily devolve into withdrawal and crying for help, and from there one can sink even further into helpless submission, despair, and the desire to surrender one's life. At each step of the way down, one weakens one's form. One's excitement becomes increasingly incapable of bounding itself—as revealed by the severe muscular clutching that conveys one's desperate attempts to grasp at the world. And what is formed is a fearful person.

The entire frustrational descent is an extension and intensification of the startled pause. Disgust, anger, helplessness, and so forth express successive stages of feeling caught in a bind. And this is one common view of the human condition: our lives are seen as a process of trying to overcome our helplessness, our anxiety. Until we do, we are victims and prisoners, terrified by the prospect of death and equally frightened of being alive. We go on feeling strong dislike for the weaknesses that we experienced during childhood; these are the same "repulsive" weaknesses we project upon old age. And we find ourselves obsessed by the need to cooperate with each other, to share what we have gone through so that we can develop insights, approaches, techniques for existence.

Another view of the human condition grows out of our ongoing experience of the formative sequence which, though momentarily checked, does not get trapped by an interfering situation. We recover from the shock of surprise. We become curious. We investigate, and either we pass on by or we allow our excitement to bloom into delighted fascination—such that we assimilate and integrate who and what we encounter.

Formative attitudes seek the fulfillment of expressing that which is newly exciting. Frustrational attitudes aim at compensatory goals. I am hungry but I'm prevented from eating, so I act too proud to eat; I say I don't care about food. It is easy to see that pride can be a forced uplifting, a stiffening of the upper torso and neck and jaw to offset established deflation and worthlessness. By contrast, the proud uplift of living a formative life comes from being filled with one's own excitement. The excitement organizes as feelings of self-esteem and as expressions of erectness and prancing.

Untying the Knot

It is important to appreciate the fact that mental attitudes and body attitudes are identical—as Nina Bull pointed out in *The Attitude Theory of Emotions.* With most educational and psychotherapeutic approaches, the mind is affected while the body maintains nearly the same frustrational form. Everybody can think of someone who, though marvelously perceptive, is still walking around with a constricted chest which gives him feelings of low self-esteem.

If a person is stuck in a frustrational rut, the trick is to bring him, bodily, to re-experience the startled state with which his conflict originated. The startle pattern is the essential attitude that needs to be touched. Then the person can begin to re-experience and investigate the re-forming of what used to be terrifying or taboo.

I begin to undo my frustrational attitudes by experiencing and perceiving them as my body. A chronic muscular contraction is not something that somebody else is doing to me. It is something that I am doing to myself. In coming to recognize how I hold myself, I begin to contact the feelings, thoughts, and memories that go along with my structure. I begin to experience the personal history of the bind in which I have bounded myself. And one way and another I make connection with the living body that I am.*

If I have developed attitudes of being a mental person, then I try to feel how I funnel energy to my head. I also take note of how I manage to quiet the excitement in the rest of my body. I experience the form of me that nourishes my thinking at the expense of my feeling and action—the form that has enabled my brain to grow while letting the rest of me become a poor cousin whose existence is alternately ignored and denigrated. As soon as I choose to inhabit the neglected part of my body, I begin the process of integrating its mes-

* Wilhelm Reich published these same discoveries in 1933, although I came to them independently.

sages: its desires and complaints. I begin to join my head with that which bears it aloft.

Attitudes of helplessness and despair are usually manifested in the chest and shoulders, since it is from this area that feelings of confidence, expansiveness, laughing, and love reach out for expression. We restrain these feelings by collapsing or rigidifying the chest, restricting breathing and crying and yelling and the sensations of softness and vulnerability. By encouraging our breathing and our self-extending, both physically and mentally, we begin to re-form these attitudes.

The de-structuring of attitudes is a natural part of the formative process. It is the part of our formativeness which opens the way to our re-forming. Though superficially they are similar, there is a vast difference between frustrational disintegration and formative unbounding. Continued frustration leads to increasingly negative feelings and thoughts, increasingly angry, restrictive, and depressive ways of maintaining one's world. The experience of formative undoing is quite another event. In learning to let go of an attitude I have held for a long time, I may feel some helplessness. I may suffer pain and mental discomfort. I may experience uneasiness and unknowing. But I also feel the flowing excitement of reorganizing myself. It's the creative void, not the pit of hell. It's my place of inner listening and alertness, not the place of inner misery and deadness. It's my place of silence in which I assimilate the old and form the new—not the place of morbid, brooding resentment that feeds on the old and aborts the new.

My process of dissolving attitudes and re-forming them is a process of learning by experiencing. If I have habitually stiffened my jaw and raised my shoulders, I experience the feeling of in-forming as well as re-forming myself in the course of breaking up these attitudes. I learn the forms of the old frustrational patterns, and I learn the forms of the new self-fulfilling patterns.

The in-forming of my cognitive sphere is called insight. Here again, what deserves attention is the *attitude* of my insight—its form, its organization, not just its content. New forming is accompanied by the feelings and sensations of mental and muscular reorganization. The experience of my "Aha!" is not complete unless this process of learning and forming involves my whole self, not simply my cognitive function. I move differently, I behave differently and feel differently in addition to thinking differently. My responsiveness to new situations keeps forming my body, my me, into being some-body.

FORM AND CHARACTER

How I Appear and How I Act

The body cannot lie. It is incapable of lying. Only what comes out of the mouth can lie; the body never lies.

My particular bodily form, my particular body feeling, is testimony to my particular character, my particular way of behaving, both psychologically and physically. Who I am has a quality that permeates every aspect of my existence and makes me recognizable. I do things seriously. I respond flamboyantly. I radiate joy. I ooze poison. This is the self that I have formed and the individuality that I radiate.

I am not so much interested in the motives behind a person's behavior. What I'm interested in is the quality with which that person performs an act—with love or with hate, straightforwardly engaged or ambivalently. I ask myself: How does this person's body reach out for contact? Is it with the groveling humility of a beaten dog? Does he, like a slave, move cautiously so as to please by his performance? Or is his quality that of a fearful defiance, a withholding by means of rigid chest and gritted teeth? Does he reach out aggressively, with stiffened arms and thoughts of revenge or tearing up? Or does he express his overall satisfaction in his tissue, which swells outward with pleasure? This is how I read character.

I respond to the process of interacting with others, and I form my self according to my experiences. I have embodied my encounters with the world, and they have left their mark. My character reveals the quality of my living experience—be it bitterness or pleasure, sulkiness or optimism. People recognize me by this essential quality. And I recognize myself by it. I have one friend who is a bouncy, aggressive taker-over, loved for his quality of sweetness; and another friend who is a slow-moving, heavy-in-mood woman, loved for her radiation of fairness.

• • •

In our early years we begin to encounter the world, and our bodily responses form our character and our awakening consciousness. There are many varieties and levels of encounter, many different ways in which life's excitatory processes may intermingle to evoke form. Body form includes the shape of feeling.

The excitement of the child and the excitement of the parent have qualities such as sweetness or resentment that communicated with each other. The communication of this quality of excitement gives the child's form an emotional tone. If the parent's excitement is heavy with over-concern or fear, it compresses the excitement of the child; the child shrinks, gives up, develops a collapsed form. If the parent's excitement is weak, spongy, unsure, wishy-washy, it provides no boundaries for the child's excitement; the child's form then doesn't know its own limits. He becomes a person who is always testing the world in an effort to find someone that will contain him.

When a child's early encounters with the outer world are restrictive—when he is almost invariably told "Don't touch" and punished for touching—he begins to be a form that curbs its own impulses to touch. A growing child who is taught to be ashamed about his sex life may come to express that shame in the form of a stiff neck. He becomes a stiff-necked character. A young lady I worked with said that when she masturbated she would stiffen her neck and clench her jaw so as to not make sounds. Another told me she would make her neck rigid when she felt sexual so as not to let others know.

On the other hand, a child whose impulses are largely accepted by the world is likely to develop a form that manifests this approval with a quality of ongoingness, pleasure, or sureness.

There are many aspects of form. One aspect of form is the shape and boundary of a cell, the shape and boundary of an organism. But form is also the shape and boundary of a gesture. There are forms of handshaking and lovemaking. There are shapes and boundaries to behavior. There are patterns of social protocol. There is the nodding between people in conversation that they never transcend. In this sense, form is the link between the visible and the invisible, between the act and the feeling of the act.

• • •

Form is slowed-down process. Part of our formative process is to build new forms that manifest the feeling of our living.

If we experience life as process, we can see that people's experience of space and time is expressed by how they are in the world. We can look at a compulsively rigid type and understand that this person repeatedly constricts his time—he's punctual, he makes deadlines and constricts his life space by constricting his body. His body is indicative of one who feels that he must not let his time and space expand, get out of control. Conversely, we take note of a weak character who cannot organize a cohesive form to contain his space and time. Lacking self-containment, self-comprehension, he leaks out, drains away. The constricted person may radiate a quality of deadness, pain; the weak person, a quality of mushiness, unreliability.

When I'm working with a person, I try to understand how he relates to his living space, how he's feeling the space which is his body. I try to perceive how much space he takes for himself, where he sets his limits and to what extent he'll allow these boundaries to be invaded. I try to develop a feel for how he experiences and lives his time, how involved he is in living his own rhythms. Each person's life space and life time is expressed as his body.

Our self-forming does not unfold in space; it forms its own space. We are not living *in* time and space. We *are* living time; we *are* living space. To possess ourselves, to be self-possessed, is to inhabit the space and to live the time that we have formed.

• • •

It is important to know not only what you do but how you do it. To know that you rebel, or even why you rebel, is incomplete. To experience the how—for example, the pushing away with arms and shoulders—deepens and completes the feeling of yourself.

There is no pat answer to the question of why you take on such-and-such a character. Why choose to be a chronically contracted character instead of choosing to be a free character? The door is never closed. The forming of character is rooted in the same open-ended process that forms us as bodies.

I remember a group situation I was involved in. One of the participants, a woman of fifty-five or sixty, had the character of a complainer. She complained that she wasn't able to complete anything, that she never felt satisfied. Her body was stooped over and sour-looking. She had chosen to form a crippled self.

She told us a dream, and it was evident from the dream that she wanted to die. Several of us pointed this out. At first she couldn't recognize it herself; but then, step by step, it became clear to her that death was what she really desired. She felt that life was a bore and that dying was something she could do successfully. To want to die was in character for her—a true-to-form expression of her crushed self. And as she began to understand how she had originated this character, as she began to understand that she was choosing to live her past in the present, she began to want to create a different form of life.

However negative a person's structure may presently appear, there was a time when it served the desirable and useful purpose of safeguarding his identity. But in terms of the person's fulfilling his potential, this once-desirable form may now be severely restrictive. And if the person maintains it too long, he dies. He dies in the sense that he effectively curtails his life expression.

I remember a woman I worked with: pretty, big-boned, energetic, a dancer, very muscle-bound, warm, with big black eyes that invited you to speak. At thirty, she was everyone's unmarried older sister. When I pointed out the big-sister role with which she approached all tasks—the attitude she had of being the non-aggressive, self-denying sympathizer—she broke down and cried. She told me that her big muscles encased her anger toward those she felt did not appreciate her or give her recognition for wanting to please. I pointed out that her muscle-boundedness also stopped her from moving like a woman. She cried again, saying she felt impelled to be a dancer so as to learn to move gracefully; she recalled that in her teens her young sexuality was laughed at by her family, who told her, "Don't wiggle like that." And we could both see how her character, the big sister, had created a shape that expressed her self-denial.

IDENTITY AND THE FORMATIVE PROCESS

Our bodily experiencing gives rise to a continuity of feeling that shapes itself as us. The shape of our own experience is our own identity. When we diminish our bodily experiencing, we subject ourselves to letting others tell us who we are ("You're a salesman"; "You're an engineer") and who to be ("Be a more pleasant person"; "Be a loyal worker"). It is our somatic messages that help us weather the need for approval and the pain of rejection. As soon as we dismiss these messages, we begin to adopt ready-made images and roles.

For years we have been content to relate to ourselves and to each other on the basis of roles that have been handed down to us. We have perpetuated precise categories for defining the nature of who we are—for pinpointing what it is to be a woman, what it is to be a man, what it is to be sexual, to be adult and mature. Most of us attempt to create our identities by imitating and living out these preconceived roles.

Ever since anyone can remember, our energy has

gone almost entirely into winning food, shelter, and safety, and our identities have been consistent with our needs. Nowadays, however, for a lot of us these basic needs are being met. And the moment our needs are met, we are presented with a surplus of energy. We then experience a yearning to move into new situations, to make new connections, to allow new forms and images to emerge in our living.

The uniqueness of us human animals is that we are open-ended. Our lifetimes continually offer us fresh possibilities for forming unprecedented relationships with others and with our surround. Our open-endedness is intrinsic to our human unfolding.

Our continual forming gives rise to feelings of joy—and also to feelings of unsureness, feelings of anxiety. When we feel anxious, many of us try to contain our excitement by holding back, holding on to the status quo, contracting and constricting rather than choosing to expand. We formulate philosophies, dogmas, and other image-systems which, by stereotyping us, assure us a sense of perpetual identity. But when we try to systematize ourselves and this systematizing is not necessary for survival, we feel shame and guilt—the guilt of shrinking away from our potential for uniquely forming our selves.

We are situated between our process of ongoingness and our attempts to retreat from that process. We are situated between wanting to establish a set of givens and being willing to let the old dissolve while the new is formed. We are situated between self-maintaining and self-forming, between preserving a status quo and moving toward the ambiguous, the uncertain.

When we challenge old patterns—stereotyped roles,

chronically rigid muscular responses, outdated feelings —we experience the present pain of our lived past. We also confront the unsureness of the future. And if we are unable to commit our energy to an uncharted course of forming the new, we retreat to the numbing security of what is familiar.

Evolution is the e-motion of living. At this particular time, the forms of our culture are slowly beginning to give way. The stereotypes are beginning to be outgrown. With the dismantling of the old and the uncertainty of the new, many feel distress and panic; a few are feeling excitement and anticipation.

Those of us who experience panic have not been able to identify with our own forming, with the feeling and pulsating of our bodies. We have accepted others' images of ourselves; we have chosen to disassociate ourselves from our excitement—perhaps because of a lack of body contact, perhaps out of despair over feeling at all. Living with and through our hurt and helplessness helps us continue to inhabit our bodies, brings us into touch with what will give satisfaction to our bodies, not merely to our minds. When we can experience our bodies' forming with pleasure, we discover our own identity.

The Role of the Role

We ask our children to find out who they are quite early in life. This happens in at least three ways.

First, right from the beginning, we expect a child to identify with a certain self-mastery. "Don't cry." "Don't touch." "Control your bowels." "Swallow that anger." There are definite roles that the self-mastered child is directed to assume: the good boy, the good girl, the obedient one, the smart one, the cooperative one. The child who takes on a social role learns to be his own criticizer in place of his parent. "I was not good today; I didn't live up to the ideal."

Second, we ask the young person to take an occupational role. "How much schooling do you have?" "What kind of a job do you have?" "What's your earning capacity?"

And then we ask everyone to take on a biological or sexual role, to identify with a particular notion of what a woman or a man is: mother, father, wife, husband.

All of these are cultural roles that shape the identity of the so-called civilized human being. In these three ways and more, one is expected as soon as possible to find a suitable niche in life. I'm intuitive, so I study to be an artist. I'm mechanically talented, so I study to be a technician. In every case there is tremendous pressure upon me to create an identifiable set of attitudes. And to create a set of attitudes is to create not only a belief system about who I am, but a set of action patterns to implement these beliefs.

A role serves two purposes. It serves to give me a shape—the kind of identity by which the world can categorize me and judge me. It also serves to give me a sense of inner continuity and self-recognition upon which I can act. By perpetuating certain definite feelings and images, I diminish my fear of the unknown. If my role is disrupted, I get disoriented. I don't know who I am. I'm out of phase with consensus time.

In order to possess an identity that gives both an outside reference to others and an inner sense of identity, most of us accept stabilized roles. I accept a stabilized identity and I die living up to it. I begin to live an approved pattern, and I cling to it all my life. The only variations are that I'm young, then I'm middle-aged, and finally I'm old. During my youthful stage and during middle age I perform as I should. And then in old age I run down—and I fear that. Whatever our age, we are all afraid of being useless, of not having a useful role to fulfill, a useful identity to keep up. So when I select my role, I feel bound to it because I'm afraid that, without it, I'll lose my connections to my self and to others.

The main alternative to the stereotypes of our Western culture has been the Eastern cultural philosophy that personality is an illusion, that differentiation from the cosmos is an illusion and that each of us is everything. Like the Western tradition, the Eastern tradition has served people as a guide to living their lives. But the Eastern approach has turned out to be no more of an impetus to growth than the Western approach. For if one is already everything, how can one become somebody?

What I'm saying is that I do not have a fixed role, nor am I everything in the world. I don't have to be a fixed thing and I don't have to be everything. I'm always forming, expressing that which shapes me, that which gives me an identity.

My formative process is revealed by all the characters, all the different physiological and psychological states of being that are manifested as my self. I may choose to identify with any one aspect of my process and live it out. I may choose to identify with various aspects concurrently—or sequentially, as they emerge. Each manifestation of my process has its own lifetime, which comes and goes. There are old manifestations and new manifestations. Some are continuous, some are discontinuous.

I am continuous and discontinuous. I never finish forming, even if I choose not to form. Steeped in wanting to be always young, I may choose to be a perpetual adolescent. I choose to be young, so I become a young old fool or a foolish old youth. In this way, my shaping process says who I am.

• • •

A number of years ago, I came to the understanding that the roles I had adopted in the course of my life were incorporated in me muscularly. I understood that in order to be a good boy, I restricted myself. In order to be smart, I coerced myself. In order to be pleasantly unobtrusive, I needed to translate unobtrusiveness into a pattern of acting. In order to hide unacceptable feelings—anger, fear, tenderness, envy, lustiness—I covered them with a tight musculature that held my breathing in and kept my neck stiff. By contracting the muscles of my limbs and diaphragm and brain and heart and digestive organs, I created an acceptable self.

The roles I assumed formed my feelings. After all, if I was feeling excited and noisy I had to suppress calling out and singing in order to be a nice quiet boy. "Being good" meant constraining my urges if they threatened other people. And so, as I formed the role of being good, I sacrificed certain feelings of aliveness from my living. I no longer knew with my whole organism that these feelings existed. My body surrendered its knowing of these feelings. They became barely a memory.

By the time I began to challenge my orderly and

obedient roles, I had begun to understand not only that they were socially given but also that *I* chose to embody them. I was accepting conditions that diminished my own resonating, my own feeling life. And when I began to loosen the muscular contractions in myself, when I allowed the bodily role-patterns to come apart, I began the excitement of feeling obstreperous and rebellious and not wanting to please, the streaming sensations of being noisy and free and sensually and sexually alive. I even began to sing again.

All in all, I was beginning to inhabit my bodily self. And the more I did so, the more I connected with the willingness to live my own image.

Roles have a positive aspect insofar as they organize collective behavior. There is no family, no agricultural community, no industrial nation without the creating of roles. But we imitate a role. Self-images, on the other hand, grow out of our individual living process. Our unique livingness initiates each of our self-images. Each self-image reflects our unique self-forming.

The Unobstructed Image

How a child experiences himself gives him his identity and generates his image of himself. This self-experiencing is greatly influenced by the communications he receives verbally and non-verbally from his environment. The environment sends messages that help to define who the child is, as well as the world the child is in. For example, a child born amid the asphalt and concrete of a city will experience nature as somewhat alien, because of being imprinted with the patterns of his artificial surroundings. He will identify with civilization as natural. A child raised in the country experiences the city as alien. His nature is not in man's works. A child in Tibet gets one identity. A child in New York gets another. The child in the mountains of Tibet identifies with being close to heaven, while the New York child identifies with being smart and knowing how to get along in the urban jungle.

When you put your hand on somebody—to push them, to hurt them, to relax them, to soothe them, to love them—you are transmitting an experience. Your self resonates with the other, and you transmit the vibrancy of your experience which helps form the other's identity. The mischief we often make in bringing up children results not only from what we tell the child; it's also the result of our mishandling the child. We may say, "I love you," but we may communicate a very different experience by saying it with unfeeling hands and unyielding, rejecting bodies. The child then feels: "I am unloved. Something is wrong with me." And the roles of victim and blamer are born, rather than the role of loved one.

As a child develops, his patterns are affected by people who are more experienced, who have more energy; and he's affected in other ways by people who are less charged. In this manner, experience is passed on. Sometimes it gets passed on as a command: "Don't touch that," or "Touch it like this." More often the identity is formed with feelings. The father's sadness teaches the child the world is unhappy. The mother's anxiety teaches the child the world is dangerous or at least uncertain. The father says non-verbally, "Don't try to be smarter than your father"; so one identifies himself as dumber than Papa, or his admirer, or he is rebelliously smarter.

Adult patterning sets the tone for the enhancing or the squashing of a child's excitement. The grown person provides feedback for how the child experiences himself, feedback which encourages or discourages the developing of a self. An adult who resonates with a child's excitement magnifies it for him and helps him to form a self-image. An adult who treats a child like an animal to be tamed and trained cripples that child's excitement. When a child has to deaden his vibrancy in order to win approval, he becomes someone who requires that others define him. The deadening of himself is the prelude to his obediently accepting and adapting to roles.

Many people have told me that their parents were unresponsive, and that this made them feel abandoned; or that their parents were not available, and that this made them feel unwanted. As children, these people identified their roles through their parents' bodies. Our impulses of love, our ideas, lose their vibrant thrust when authorities respond negatively, or when they don't respond at all. Sometimes their disapproval reinforces the tenacity of our vibrancy; but more often, especially when the authoritative *no* will not brook alternatives, we develop a fear of forming who we are.

Nowhere in the West or the East do those on top tell us that we can identify with our pulsations, our streamings, our feelings. Some authorities say that we have such things, but we are not to trust them and we are certainly not to identify with them. Nonetheless, I fully believe that the human being who denies his bodily process is denying the only identity he can ever go on having. The lived body which feels and dreams is always forming an image. When we take away all the roles—doctor, lawyer, Indian chief—our biological experience gives us a personal vision of continuity and connectedness. This is our living image. By identifying with it, and by maintaining contact with our bodily experiencing as the source of our self-referral, we develop a sureness, a soundness, a faith and pleasurableness in the forming of our lives.

SELF-FORMATION THROUGH DENIAL

Self-formation can be articulated by chronic muscular contractions, which tend to deform our thoughts and emotions as well as the gracefulness of our bodies. If I chronically contract my chest, I feel unloving and unlovable, and I believe that life has ill-treated me, passed me by. If I consistently restrain my crying and clench my fists to keep from striking out, I smolder with unexpressed vexation and offer the tight-lipped opinion that it's a dog-eat-dog world.

To contract one's self is to restrict one's frame of reference. Try it: close your eyes and tighten your neck. Hold your breath; squeeze your shoulders together. And now feel how you are inside. Experience your space. What kind of picture of the world do you get? Now tighten up again and make a sound. Listen to your contracted sound. Then, with your eyes still closed, release your squeezing. What happens to your sound? Open your eyes and greet the world. To release a muscular contraction is to permit the world another face.

A muscular contraction has positive value insofar as it serves as a short-term personal defense. To contract is our bodily way of saying *no* to interrelating fully with our selves and with others. We tighten up to prevent getting hurt. We shrink away from bodily harm, whether the jeopardy arises from outside or from the impetus of our own feelings and needs. To diminish our selves is to diminish our sensations. In restricting ourselves, we restrict our sense of pain—and of pleasure.

A child risks death if he does not or cannot respond to the inhibitory vocalizations and gestures from his elders. If he's skirting the edge of danger, it's life-supportive for him to react to his parents' "No!" But when his parents arbitrarily use their *no* to the extent that inhibition becomes training, there begins a war of wills. The child has a *no* too. And if the parents' *no* repeatedly overwhelms the child's *no*, the child begins to feel meddled with. Bit by bit he develops the picture of an antagonistic world: a world in which it is unsafe to move, a world in which exploration brings about catastrophe.

When a child says *no* he means it. *No* is a statement of selfhood. It weakens a growing person if he does not come to value and utilize his *no*. And if the parents abuse their *no*, the child has a very hard time learning to use his *no* without similarly abusing it. Any parent who harps on *no* for the sake of discipline or constant protection is ignoring the *no* of the child. And the child's response, in addition to undervaluing his own *no*, is either to disregard or to submit to the overwhelming *no* of the parent.

If, as a parent, you feel your *no* when you communicate it to your child, he will listen to it and respect it. But if *no* becomes a habit, if it becomes bound up with a stern system of child-raising, then your reprimands dampen your child's excitement and help to form spite. The child who is regularly scolded or slapped when he talks up learns to clench his jaws, and he continues to do so long after he leaves home.

• • •

We overuse our ability to self-contract largely because it gives us a sense of power. When we narrow ourselves and restrict our pulsating, we interfere with our process. We create for ourselves the illusion of having stopped time, of having achieved a static reality. We believe we're safe in this static situation. And we believe that we are saved—that we have clinched our immortality.

The holding-on of muscular contractions produces sensations of foreverness by impeding metabolic activity. In tightening one's self, one narrows the stream of one's present; one slows down the pulsing flow of one's life and forms the fantasy of extended time—a forever-fantasy which one may visualize in terms of the past or the future. But not the present. Either way there is an avoidance of present experience, presentness.

One's state of foreverness supplants one's feeling for the truth of how life comes and goes, forms and unforms, is born and dies. It's absurd. Most people are so devoted to immortalizing their present lives that they cannot postulate a life hereafter that is substantially different from this life. They cannot and will not permit themselves to conceive a new form of living. By cramping their bodies, they deform their perceptions—their own feelings and images of the many ways to be alive.

Our Western culture has taught us to cultivate the misuse of our inhibiting abilities. Stereotyped attitudes are encouraged and consciously perpetuated. Think of how we learn to mask our faces in order to hide our feelings. Consider how we practice not to cry with our chests and bellies, how we work to squeeze our hurt into our heads, where we can most efficiently disguise it. For the past several hundred years we have been drilling ourselves in all the various techniques of constrictive power. And in so doing, we have smothered our passion and natural empathy. We have cramped our emotional selves.

The heart feelings of love, tenderness, desire are basic expressions of our vibrating, pulsating stream. When we contract our streamings in such a way as to restrict the feelings of our hearts, we become creatures who do not emotionally respond to the aliveness of our selves or of our environment. And so it is. Having trained ourselves to be emotional deserts, we are now turning our planet into a desert.

Narrowing Our Selves

As soon as a contraction ceases to be pro-survival and endures past its moment of usefulness, it becomes a negative factor. Chronic muscular contractions, whether culturally induced or self-induced, are self-defeating because they are overly self-protective, self-separative. If we maintain a constricted attitude, we become disconnected not only from the world around us but from whole segments of our own being. The irony is that to avoid danger, to save ourselves from dying, we deny part of our living.

At one time or another, we have all sacrificed parts of ourselves so that the rest of us might live. For example:

1) The man with the caved-in upper chest, shoulders bowed. His head sticks out like Quasimodo's. What is he expressing if not self-denial? Instead of being filled with his uprightness, he crushes himself. He's afraid of his excitement—afraid that it might control him if he doesn't control it.

2) The woman whose pelvis is tucked in like that of a beaten dog. Her legs are tightly pressed together; her fanny is packed away underneath. She may have a paunch. With the lower half of her body she's squeezing and pushing forward, while her top half is leaning backward as if pulling away from something distasteful. She won't receive and she won't give. She's fearful about filling and emptying. "Make me open up," she says. "I dare you."

When somebody comes to see me with hunched shoulders or a tucked-in tail, we try to find out who he has become. I also attempt to move him toward asking himself, "What is it I keep doing? What am I not letting live? And how does that feel—my not letting it live? Am I afraid of what might happen if I let my shoulders down?" And each time, the two of us come to the amazing discovery that the freedom to be responsive is frightening to this person. "If I let my shoulders down, I may express the anger I feel toward my father. I might even hit him, and that wouldn't be right because I'd become what neither of us wants. So I'd better keep on sacrificing my urge to lash out."

We choose our manner of suffering. The part of us in which we suffer is that part in which we handle intensified excitation poorly.

I remember working with a woman who had multiple sclerosis. One time a burst of feeling rushed through her legs, and she said, quite spontaneously, "You know why I crippled myself? Because whenever I went to my mother, she wasn't there." Her decision may have been unconscious, but nonetheless she *chose* not to walk. She said *no* to her expansive impulse to walk.

Negative influences from the environment acted as dampeners on this woman's unfolding excitatory patterns, contributing to her choice of infantilization and subsequent deterioration. Later on in life when she felt a resurgence of the impulse to expand, her own legs restricted her ability to do so.

Yet she was most keenly alive. All depressive people are alive; they are alive in an excruciating way. Their aliveness is unbearable to them. So many people come to work with me who cannot bear the life they have at the moment, and yet they claim that they want more living, more excitement. What a paradox!

Containing More of Our Selves

We may perceive that our contractedness is undesirable. But as long as we don't know how to contain ourselves flexibly, we're stuck with our rigid containers.

The art of living our bodily lives is that of continually creating new containers, evolving forms. When our boundaries are chronically restrictive, we resemble islands. If our boundaries are underdeveloped, we keep perpetually active in order to avoid drifting into fragments.

To contain ourselves is to embrace our living without hugging it to death. Signs of containment are a fullness of feeling, a deepening and ripening of self-experience. Think of a full stomach, or a breast full of feeling. At what point does containment stop and self-denial begin? Think of a distended bladder, or a breast engorged with milk. When we reach the edge of expression, the excitatory apex of our formative process, we hold back. We won't let the next step happen. We won't give up our container. We tighten, we defend ourselves. We register as doubt and pain this preventing of the natural expression of our contained energies. And if we continue to hold on, we experience an overall diminution of feeling and excitement. We compress ourselves to death.

In the course of my living, I build temporary boundaries. I set transitory limits for my self. A stiff neck might be the boundary that expresses my present feeling of "Don't invade me; I'm uneasy just now." It is impossible for me to be alive without defenses. It is impossible for me to risk interaction without inhibiting and limiting myself to a greater or lesser extent. Living does make me hurt sometimes. Interrelating with the world does lead to my erecting protective barriers. To experience my personal inadequacies and work them through inevitably leaves its mark. Each inhibitory event forms my character, like the gnarls in a tree.

But I also wish to open up my boundaries, to move beyond my inadequacies of yesterday. To be open to my living process is to feel my willingness to keep on experiencing. And so I keep coming to the place where I ask myself, verbally or non-verbally, how much I am willing and able to experience. Sometimes I find that I've forgotten the key that unlocks my old boundaries, and I ask a locksmith to help me—either to fashion a new key or to pick the lock.

Yet even with the help of a locksmith, it's not enough that I unlock myself. It's not enough that I simply stop denying myself. I need also to begin learning how to affirm myself. This is not brain learning; it's learning that takes place throughout all the tissues and organs and fibers of my flesh. I learn with my body to swell with my excitement, to contain my streamings, my feelings, my thoughts and perceptions. And this is the thrust of my life.

THE DECISION TO FORM ONE'S GROUND

The following was transcribed during a weekend seminar for professionals, a teaching situation. Participants made comments, and there was an explicit attempt to weave theory into practice, to show how the self learns to perceive itself and make self-corrections, experiencing its illusions and coming to be more vividly in the present.

When people come to me for help, whether individually or in a group, I usually ask them to get partially undressed so that both of us can see as much of their body as possible without invading their privacy. In most cases they present a life problem, and we try to tie up the nature of their life situation with the form and movement of their body. In this case, Fred volunteered as a subject to be diagnosed—to see if we could deduce from his bodily stance what his problems might be. As it happened, the diagnostic procedure turned into a healing process.

Fred was tall, big-boned, and he held himself stiffly.

He looked like a boy who had worked at muscle-building to make himself into a man, a boy who was trying to appear bigger and tougher than he actually was. His facial expression was that of a statue—mask-like, stony, unrevealing. He wavered on his feet like a flagpole in the breeze, ill at ease, uncomfortable with his ground. To me he demonstrated three distinguishing characteristics: his boyishness, his inflatedness, and his stiffness. I was interested to discover how these bodily characteristics defined the life situations he got himself into.

• • •

Stanley: The first question I'm asking myself is: How is this person grounded? What is his relationship with the earth? What is the quality of this connection? How much energy does this person present as he stands here? Is he dull? Is he vibrant?

What is his shape, emotionally and physically? How is he embodied? How does he inhabit his flesh? What is his body trying to say? What kind of feelings does he communicate to you? What is his living statement to you?

Participant: I see more aliveness around his head than around his feet and legs. I see an egghead.

Stanley: Okay, what is the feeling of this egghead image? What feeling is this person organized around? Is he organized around hopefulness? Around despair?

Participant: I feel there is a sullenness.

Participant: . . . a remoteness.

Stanley: Then I would ask him: What is that sullen look? And how is that sullenness related to his egg-head, to his cerebralized state? Why is he brooding? What is he being? I experience his way of being in the world, and how he feels to me helps me begin to make sense of him. His sullenness tells me how he is grounded and gives clues to why he doesn't want to become more grounded—why he stays up in the head a lot.

Participant: Could you explain again how he begins to make sense?

Stanley: His sullenness communicates a feeling of disappointment. Somewhere in his living, he didn't get something crucial that he needed. We have to find out what was unfulfilled in him. Now he expresses a silent demand. Is it "touch me"? Is it "hold me"? What is it? With the sullenness there is a quality of reserve, of resignation and withdrawal, causing a bodily density and a shrinking from the world.

He has vitality, too. One sees that in his eyes. This person is strong. His body is not weak; it's got substance. His tissue is not like pudding. Maybe he's ungrounded because he's caught between wanting to connect with us in a vital way and avoiding the possible disappointment of being turned down.

He keeps his shoulders drawn close to him, cramping his chest. What does this do to his life space? What kind of feeling comes out of this constriction? What kinds of thoughts emerge from this restricted ability to expand?

Participant: What I get as I look at him is a kind of statuesque quality.

Fred: I don't know about "statuesque." I'm standing here trying to really be here and at the same time listen.

Stanley: Okay, I agree with you, but how are you there? You look to me as if you're sitting behind a desk in a classroom—obedient and immobile.

Participant: You remember, Fred, yesterday I said to you that I thought you had difficulty feeling. And yet you're extremely perceptive. You pick up all kinds of things in everybody else. You're the first one to pick up on things, and you're right about them. And I'm wondering about the contrast.

Stanley: How does his bodily shape express that?

Participant: He's withdrawn into himself. He's standing there like a watchtower.

Participant: Is his body saying that to move is to lose some of his perceptiveness of the environment?

Stanley: Or is it saying that if he moves, he's afraid he'll lose his sense of himself?

Participant: To move is to feel.

Stanley: Right. So he diminishes his movement. He pulls himself off the ground.

Let's broaden this. How do you think he makes decisions? How do you think he exercises his freedom, his range of potentialities?

Participant: When you first stood up, Fred, it came to mind that you seem to be caught between resignation and defiance, and that's what makes the immobility. Neither one of those really gets expressed.

Stanley: That's important. He's living the impasse.

Participant: Fred, I was noticing that when people are speaking to you, your eyes seem to pull back. The slits narrow as if you were peering from behind a thick wall.

Stanley: Okay, that's enough negativity. You can't appreciate a person by staying only with what's wrong. See if you can discover what's positive about him.

Participant: The body is well proportioned.

Stanley: Is his mind well proportioned?

Participant: Sort of a courage.

Stanley: Is he brave in his perceptions?

Participant: There's a certain integrity.

Participant: A directness in the eyes.

Stanley: Yes, my feeling is that this man has indeed made decisions of directness. He has made the decision to be truthful, and this decision is expressed in his watchtower stance and by his sullenness, which he doesn't try to hide.

Participant: The right word seems to be somewhere between integrity and self-sufficiency. He doesn't need to take from anybody else to maintain his own thing.

Fred: When I was a teen-ager I decided there were two ways of being big. One was to beat everybody else down—

Stanley: Fred, are you aware how your face was just gesturing?

Fred: —and I decided it's not right to try to beat everybody else down.

Stanley: What are you moving toward?

Fred: I'm not feeling as resigned. I feel softer, and I feel a trembling in my body. I feel I'm not struggling with myself to hear what you say without cringing from the impact of it. I feel I can be open to it in spite of the fact that it's going to hurt.

Participant: His eyes are less piercing.

Stanley: How does that less-piercing quality feel to you, Fred? How are you experiencing it?

Fred: More sensation in my eyes, and my mouth opens a bit more.

Stanley: Is this being vulnerable?

Fred: I'm not sure. . . . I feel a kind of delicious humiliation. Then the shame of the deliciousness. In my eyes I'm proud. And sad.

Stanley: Your expression looks like "Don't hurt me."

Fred: I'm not sure what's going to happen from inside me. I both want it and don't want it.

Participant: I hear a martyr-like quality in his voice.

Stanley: He doesn't really appreciate his own strength. I'd say that he has sacrificed some of it in order to protect himself against the possibility of getting hurt. That's the watchtower and the piercing eyes.

Fred: That's what I've been reaching for. Not to be afraid.

Stanley: Try hitting the bed. Breathe more with your chest. Every bit of assertiveness that you allow to emerge is you. You can be as assertive as you want to.

Fred: [Slowly crumples to the floor and starts to laugh.]

Stanley: What's funny?

Participant: His watchtower finally collapsed, and nothing tragic happened. If you're thinking something terrible is going to happen, and it doesn't, then you laugh.

Fred: [Starts to cry.]

Participant: That whole sequence was one of the most unusual things I've ever seen. Really, Fred. It began with your statement that there are two ways of being big. You told us one way: beat everybody else down. But you never got around to the alternative to that, which is: "They're not going to beat *me* down." And that's the defense that you have adopted.

Stanley: I also saw a willingness to be defenseless, to let your tautness slacken, to give up form, to learn to trust what you were forming at the moment, to let yourself take a different stance in the world.

To learn is to let go of the trumped-up performing that comes out of your upbringing. To learn is to accept an experience which you have not yet coordinated and synthesized—to accept this experience without resorting to models of how to perform. In the realm of feeling, behavior is invented rather than imitated.

Fred, as assertiveness becomes more available to you, you'll begin to integrate and shape these feelings, and you'll find expression for them. They'll give you form.

Fred: It's just that I'm aloof, shy. I hold back. As a therapist I work at allowing myself to feel my patient's pain without having to rush in and stop it. We had some guests over, and their little two-year-old was knocked down by our dog. The head hit the concrete, and if you live in the city you know what that feels like, and my brain was jarred and I realized what I'd been fighting against. Who wants to feel someone else's pain?

Stanley: Say that again.

Fred: Who wants to feel someone else's pain? I feel *my* pain. That's okay, because I can accept it as mine and do something about it.

Stanley: Many of us can't differentiate between our own pain and the pain felt by the other members of our family. Was this true for you?

Fred: My mother was depressed. She tried to live through me. She whined and nagged and was chronically dissatisfied. I did everything to keep away from her. You know that scene in *The Graduate* at the breakfast table, where he's reading the cereal box?

Stanley: Okay, you were shaped by your mother's pain. But what was your own tending toward, your own wish?

Fred: I wanted to be what she wanted me to be.

Stanley: Did you ever experience this as painful?

Fred: I did.

Stanley: How about right now?

Fred: I don't think I want to feel it.

Participant: Maybe I'm caught in my own projection, but I saw a lot of masochism there. I saw it in the crumbling.

Stanley: What happened here may have looked like crumbling, but it was not. He gave in, which is not a collapse. And it isn't masochism; it isn't self-defeating. Giving up form is not wallowing in the pain and suffering and making a life out of it. Fred allowed himself to become unorganized, and out of this came a new form, a new self-assertiveness. He never collapsed like a house of cards.

A person who can give love and receive love is self-expanding and self-affirming. When Fred began to experience that his loving wasn't received and that his mother's loving was conditional, he formed his watchtower, his remoteness and sullenness.

Fred: Yeah, I was damned if I did and damned if I didn't. I felt trapped. If I loved her, I became her slave. And if I allowed her to love me, I lost my identity. All I could do was pull away and go stiff—hide my need and ward hers off. When I fell on the floor I could actually feel that stiffness breaking up. And all of a sudden I started feeling warm without feeling scared. I felt myself having a shape without being stiff.

THE UNCHANGING BODY

Many people say that the body doesn't change, that it just grows older. Many people feel that they cannot change. Others feel that they don't know how. Still others refuse to change. The inability or the unwillingness to change permits a multiplicity of possibilities to remain unlived.

Roberta was one of a group of people trying to discover how their unlived desires interrelated with their fear of dying. She complained that she never got what she wanted, which was to be loved but also to be independent and free. She could not sort out how to be loved and still be her own person. This left her lonely.

She considered herself smart, smarter than most men, and she claimed to be misunderstood by them. She also claimed that men treated her as a sex object, and yet she dressed and acted provocatively.

Roberta was about five feet five, dark-haired and trim. Her bones were of moderate size and she had good muscular tone. She tended to be stoop-shouldered, with contractions in her chest and throat, and had a permanent pout planted on her puss. The pout was testimony to her self-indulgence, while her constrictedness indicated her diminished willingness to love and to be loved. She presented a quality of hardness in everything she did.

• • •

Stanley: What's going on with you, Roberta?

Roberta: I just feel dead.

Stanley: You haven't looked well, either yesterday or today.

Roberta: I feel good in the morning. A lot of shit is pouring out.

Stanley: As if you were exploding. And yet you look to me more as if you're shrinking.

Roberta: That's what happened last night, with a friend of mine. I shrank away from him. I couldn't let myself—

Stanley: —"wouldn't" rather than "couldn't."

Roberta: Yeah, I wouldn't. . . . Anyway, later on last night, I had this dream about two men on a mountaintop. They were looking down on a woman who was falling into quicksand and sinking. It was almost like a naval burial, watching her sink.

Stanley: What kinds of feelings did you have along with that?

Roberta: There was a kind of hopeless quality to it. Like the guys on the mountaintop were . . . impersonal.

Stanley: Was it claustrophobic, self-shrinking?

Roberta: Yesterday was frightening.

Stanley: What was frightening about it? What was the shape of that fright?

Roberta: Well, it started off when you were doing your theories and stuff. I was getting angry because you were talking too fast.

Stanley: Too fast?

Roberta: It was almost as if you didn't want anyone else to interact. I felt slighted because there were a lot of things I wanted to say. And I got into my feeling of being worthless.

Stanley: Is that like not being recognized? Like being small?

Roberta: To me they're the same thing.

Stanley: Do you want to say something about feeling unrecognized?

Roberta: Oh, like last night I was thinking that I wouldn't be feeling this way, I wouldn't be so upset except for listening to this tremendous urge in me to be recognized, and knowing that I could never get enough. And so it's a vicious circle. There is a vicious quality to it.

Stanley: Why do you use that adjective?

Roberta: That's what I feel it is at times. I experience it as that.

Stanley: As vicious.

Roberta: Yeah.

Stanley: How does vicious feel? What's the form, the shape of vicious?

Roberta: There's a squeezing.

Stanley: Squeezing is vicious?

Roberta: It can be vicious.

Stanley: Yes, it can be. But you haven't really described your own style of viciousness yet.

Roberta: It's a feeling of . . . actually doing harm to another person. The enjoyment of it: focusing myself to dominate people.

Stanley: Being vicious serves to give you a feeling of power?

Roberta: It's like sometimes when I screw, I hold back. It makes me feel strong, hard, tough.

Stanley: That's vicious? I think holding back can be very pleasurable. In fact, one of our greatest sources of pleasure is that we can say, "Hold it, I want more. I'm not yet ready to let myself go here."

What were you telling yourself in that dream?

Roberta: I'm not sure. Usually my dreams are clear—the visual sensations, the physical sensations. But this time there was a kind of deathly gauze over the whole thing.

Stanley: You could see yourself?

Roberta: No, I didn't recognize anybody. I was seeing

the two men on the mountaintop, and the woman in the quicksand was going down. I was somewhere else.

Stanley: Who's the gal in the quicksand? Could you act the drowning person—be her body? . . . What's the experience of that part of yourself?

Roberta: I don't know, but . . .

Stanley: Roberta, what are you crying about?

Roberta: Part of me wants to live—

Stanley: —and part of you wants to die?

Roberta: I'm trying to keep any sound from coming out.

Stanley: Do you feel that anybody cares?

Roberta: I don't care if they care! I'm tough! . . . It felt good to say that.

Stanley: I wonder if it's true, though. Because the minute I began to pay attention to you, your whole attitude changed. You became softer. You stopped holding yourself so tightly. You expanded. Maybe you care so much that you can't bear the pain of it. Maybe you care too much.

Roberta, you say you want more freedom. Well, you've got a fundamental choice to make, a choice that you have to make consciously, about how you're going to respond to yourself in order to be free of your stereotypes. You can take on the traditional point of view that you've been born in original sin, the point of view that demands that you constrict yourself, that you hold the body in bondage and deny your sensual impulses. Or you can define yourself as a tender animal, full of lustiness and light. Either you confine yourself in such a way as to perpetuate the form and feelings of being tough, hard, small and tortured, or you respond to yourself as someone who wants loving and caring.

Why don't you lie down, Roberta, and see if you can locate where it is that you're grasping yourself. Put your hand on that place and see if you feel anything. And even if you don't feel anything, you can experience the deadness. What's its form? Recognize that you have chosen not to be responsive in that place, that you've allowed your form to be frozen there.

This choice is not something to be overcome. It is not something to be gotten rid of, like a piece of garbage. It's your life experience. It's *you*. Can you feel the form of the covered you, the hardened, muffled you?

Roberta: [softly] I choose to muffle myself.

Stanley: Where, Roberta? How?

Roberta: The back of my neck.

Stanley: Put your hand there. Say "I."

Roberta: I. I. . . . But I don't feel anything.

Stanley: Not even the tightness? Would you say, "I choose not to go any further, not to respond?"

Roberta: I choose not to go any further.

Stanley: What feeling does that have for you?

Roberta: It gives me the feeling of fear.

Stanley: Can you respond to your fear? Can you let it speak to you without being overwhelmed by it?

Roberta: Yes. . . . I'm feeling more relaxed now.

When I started listening, I had a very strong feeling of letting go.

Stanley: Of what?

Roberta: My need to be tough.

Stanley: You chose to receive something.

Roberta: Yeah, I felt that I was drawing something in and I didn't choose to block it out. It was a very pleasurable sensation: letting go, letting in.

Stanley: What were you letting in?

Roberta: Myself. Me. My excitement. I was aware of something happening in my chest and my stomach, and I tried to get in touch with that. I chose to let go of quite a few things, and now I'm feeling very down-to-earth and relaxed. I'm going to have to let that part of me come out.

Stanley: What part is that?

Roberta: The part that I just wouldn't ever let out. You know, like caring. I can see that I've got to take the responsibility for letting me live a different life—to live myself differently. To learn from my own urges.

ALTERNATIVES TO INTROSPECTION

ALTERNATIVES TO INTROSPECTION

I look around me and I see that everything has become something. Everything has undergone formation. In interacting with the world, everything has become more than it started out as.

A lot of forming goes on outside the purview of self-awareness. I think of myself one way, and then I discover that I'm quite something else—and this happens whether I am conscious of my self-forming or unconscious of my self-forming, whether I knowingly participate in becoming someone or unknowingly participate in it. Even when I am actively involved in shaping myself, I may suddenly realize that Hey, I've been formed! I'm me! I am the unique expression of the whole of my life, the particular quality that has emerged and established itself as me.

When I work with a person, I keep a close eye on what is emerging in terms of what he or she is presently forming. It's the kind of looking that leads me to interrupt and say, "Are you aware of what you look like? Do you experience what you're doing to yourself? Do you realize that your neck, your chest, your pelvis have begun to move and that you now have another rhythm?" In a sense, the person is learning to be an artist. One man I worked with became a poet, even though he never wrote a line. He came to experience and perceive the world as a poet. He began to speak poetry, to live the form of a poet.

Many times people undergo formative changes and they're so keenly interested in their performance or in the emotional content of their various reactions that they miss what their form has come to be. They don't really experience that their shoulders have let down, that they have unexpectedly grown more graceful, more coordinated, more radiant, that their self has enlarged and that they are literally someone else. They ignore who they have become. And so they keep on talking in the vernacular of patterns that belong to the past: "My mother is bothering me." "I can't be happy." "I can't trust my ground."

The great majority of people involved in therapy—and in education—want to know. They want to know who they are, how they relate to themselves, how they act and react. They want to create a cognitive connection between what is going on now and what has gone on in the past and what may go on in the future. But I'm not all that interested in whether or not people know. What interests me is how they are or are not *experiencing* who they are becoming—how they are or are not *experiencing* what is forming.

I encourage expression. I encourage people to let their excitatory processes speak. I ask them to feel themselves and to move different parts of themselves. I ask them to feel what emerges when they touch themselves and when I touch them. I ask people to feel their breathing as it is, and then sometimes I ask them to slow it down or speed it up. With certain people I ask that they act aggressively: that they stamp, hit, shake, kick, shout—so that they can observe their rage and decide whether to support it or be rid of it. In essence, I ask people to try new forms, to hold their bodies and to use themselves in unaccustomed ways.

The content of a person's understanding is important, but to me it's nowhere near as important as a person's being able to experience the forming of his own attitudes, his own urges, his own expressions—the myriad idiosyncrasies of being himself in the world. And therefore my work attempts to initiate a series of unfamiliar experiences which will enable that person to teach himself about his excitement and its rhythms—experiences which will enable him to feel and perceive how he contains himself and extends himself, how he relates to the known and to the unknown, how he establishes boundaries and gives up boundaries, how he releases old form and creates new form.

• • •

We expand and contract. This pulsatory process is the story of how we form ourselves, the history of our image and identity. Our dream does not have to be searched for inside. It presents itself without our rummaging around in our psychological graveyard. The dream expands upon us, into us, like a wave swelling forth from the ocean. It engages us physically and cognitively, seeking not interpretation but further understanding through self-reflection and through expression in the social world of friends and lovers. One soon learns that one's forming has a thousand tongues.

Intensifying the expansiveness and the thrust of our excitement paints a very different picture from turning back on our self. We let our inside become the outside. We let our excitement surface. In response, the thrusting world invaginates us; its exterior becomes our interior in a dance of alternating surfaces. It is a dynamic process, a process of forming rather than performing, a process of cooperating rather than competing, a process of experiencing rather than a constant effort toward introspection.

EXPERIENCING

Our formative process is mother to our experiencing, just as experience sires forming. Experience directly expresses three aspects of our formative process: (1) the quantity of excitation we are able to contain and release, (2) the qualities of this excitement (hard or gentle, weak or strong), and (3) the rhythmicity of our excitement (its ebb and flow, its contracting and expanding).

We are most globally excited when we are in the world with little boundary—when we're immersed in the expansive (pre-personal) and expressive (post-personal) phases of our formative sequence. But in these phases we have the least *sense* of experience. It is not until we contain our excitement that we gain the self-feeling and the self-perception that round out our self-experiencing. In forming a container, we alter and diminish our connectedness with the excitatory matrix, enhancing our individuation. As we draw into ourselves, we come to know more about what is going on relative to us, both outside and inside. We create a distance which permits us to reflect, to conceptualize our selves and our world. We are then in a position to connect even more lusciously with our surround.

The embodying of excitation makes self-aware experiencers of us. Containment stimulates our development as experiencers. In our unbounded phases we are not the experiencers, we're the experience. We don't *perceive*; we *are*. There is no reflective I.

As children, we express excitement impulsively. Then, as we start to develop boundaries and become capable of containing our excitation, we begin to experience: "Here am I, and there is the not-I." "This is my self, and that is the not-myself." "Here is my world, and there is the outer world." Our sense of distancing and discriminating and focusing is formed in the embodying of our excitement.

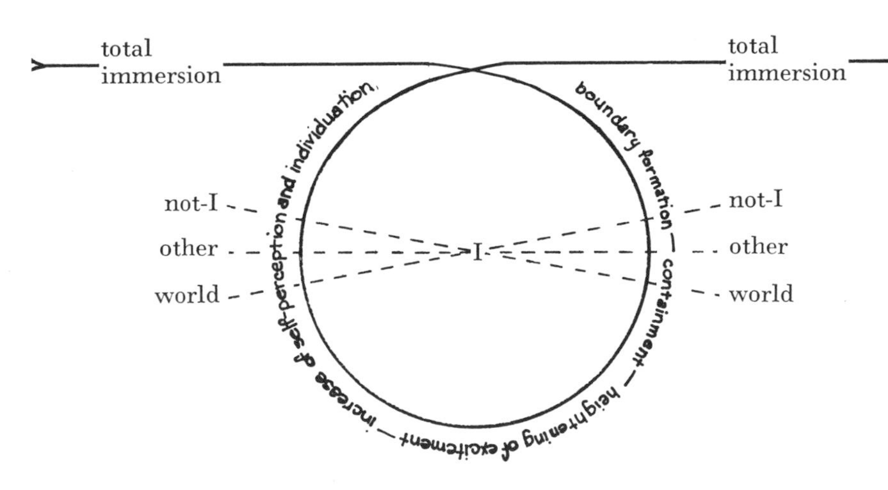

Continual immersion does not make us aware of our experience, nor does it enable us to utilize it in the ongoing process of forming ourselves. Those of us who claim to be totally involved do not fully experience that involvement until we begin to pull back and reflect. Our embodiment sets bounds, creates distinctions. I, as body, experience the not-I, become intimate with it. This intimacy of I and not-I forms a new I, a new entity which makes the former I and not-I knowable.

We have no human voice without our body as a sounding box. In the same way, without a body we do not resonate with experience. Each of us is a uniquely embodied field of oscillating excitation which resonates in the field of the biosphere. The experience of immersion is that of resonating with the basic excitement of life and with other bodily creatures in this more inclusive field.

The different frequencies of our bodily resonating give rise to the different strata of our I, which make self-reflection possible. We contain the pulsating and streaming excitement of that with which we have been able to resonate directly. Our containing allows us to absorb and digest the experience of our immersion. And then we go out again. We undo our roles and images, our boundaries. We stretch our limits, shrink our distances, and resound with the world.

Then, once more, we become willing to diminish our connectedness and to re-form. There is the immediacy of involvement, then the diminution of connection, and then re-formed connectedness. This is the pulsatory ongoingness of our formative process, in which we reach to the surround and draw back. We steep ourselves in our environment; then we separate out, assimilate, and reflect on what has taken place. This is how we nourish ourselves, how we deepen and broaden our experiencing.

Our Inner Sea

Once, flying into San Francisco over the salt flats near San Jose, I looked down and suddenly saw how a supersaturated solution begins to form a crystal. I saw a crystallizing field in the water, a field that had reached its limits of expansion and was beginning to set up borders and maintain itself. Moving with the tide, resonating harmoniously within the water, it resembled a spider's web in the breeze. And I understood, experientially, how we all are in the world . . . and how we all are the world.

The human being is a small sea of streamings in the great ocean of excitation. The streamings of each small sea present a particular resonating pattern—or rather, a particular combination of resonating patterns, superimposed one on top of another like a stack of grids. I'm like the crystalline pattern in the salt water; the world and I oscillate in unison at certain frequencies. And my resonating pattern is capable of contracting itself. As I contract myself, I slow down *part* of my oscillating field. This slowing-down forms structure, boundaries, for containing and expressing the sea that is me.

My experiences—my feelings, thoughts, and actions—are derived from the geometry of my vibrating, pulsating field. My bodily geometry determines how it is that I resonate. My experience is most intense where my field oscillates most vigorously. Think again of the human voice. It resonates in the head, the chest, the abdomen, and even throughout the entire body. Wherever it resonates, one is intensely aware.

• • •

The deep-rooted feelings that arise from experiences of being pulsatory and streaming contrast with feelings such as annoyance, itchiness, titillation, and localized sensation, which come from experiences of simple excitation. Flirtatiousness and cheerfulness, jealousy and prideful hurt are peripheral compared with the pulsatory quality of love and joy, rage and grief. We have all seen those whose tears were a shallow attempt to pass for crying or whose apparent anger we recognized as a temper tantrum. Superficial feelings do not involve the whole person, only the external layers. The difference between excitatory feelings and pulsatory feelings is the difference between a flickering flame and deep, hot coals.

The levels of our experience correlate with how thoroughly and profoundly we feel our excitation. At the deepest level, our experience is direct and immediate. At a less profound level, we interpret our experience in the form of archetypes, symbols, and dreams. At a surface level, we interpret our experience as sensations and thoughts.

The tendency in our culture, unfortunately, is not to encourage the multi-level experience of excitation. Instead, we create conditions in which the forming and unforming of excitement is restricted to the level of the skin, the brain, the sex organs. We experience the world with peripheral pleasure rather than deep satisfaction, with vexation rather than heartfelt anger.

Most of us are able to let go of surface involvements, but we have difficulty letting go to the extent that we begin to experience deeper levels of our own bodies and the bodies of others. We may understand cognitively that, at all levels, the unforming of our boundaries is necessary for expanding our experience, but it still isn't easy to let go. We experience our *selves* most intensely in a contained stage. And individually and culturally, we have generated a powerful fear of giving up the continual self-experience of containment, such that we don't dare to open our living to levels of experience that we know nothing about.

As I listen to people's experiences, I hear to what an atrocious extent we teach ourselves to deny and even denounce our own flesh and blood—thereby diminishing connection with the depth of our formativeness. Most of us form our lives superficially, while the forming of our deeper selves is neglected and discouraged. Caught up in the superficiality of our performing, most of us have very little experience of the depth of our possible forming.

Learning

We train the brain to control and discipline the body. We are brought up to reason with our instincts, to cajole feelings with thoughts, to earn gratification by being smart and doing a good job. We appoint the central nervous system master and driver of the rest of the body.

We foster the belief that cognition is the great experience. This is what we call "learning." Almost all of our social learning forms are based upon the assumption that experience and its communication are cognitive.

I'm suggesting that learning—the transfer of experience—is a culmination of the resonating of the body's excitation. The body's resonating lights up its powers of cognition. The brain is the servant of the body, not vice versa.

A friend of mine told me that he used to give himself erections by calling up pictures into his mind. He invested his excitement into this picture life. As soon as he stopped inflaming himself mentally, his body experienced more feeling and movement.

The dynamics of body-directed learning are illustrated by what happens in the formative sequence. Excitation expands, collects itself, and forms boundaries. And then, as it begins to express itself, it reaches past its boundaries—past its loop of containment and into the surround. Expression is a peak event, like the peaking of a heartbeat or the peaking of orgasm.

The formative peaking interacts with the environment, generating new in-form-ation, new communication. Then it draws back, gathers itself, and forms a new loop. The new containment invokes all of the feeling, imaging, dreaming, thinking, and decision-making which permit a richer experience of self and other, self and world. This is how we form our reality. This is how we learn.

In a well-known study, two researchers showed that when a child's creeping and crawling are interfered with, he has trouble with his speech functions and often ends up as a stutterer. Their study was a revelation insofar as it demonstrated that when certain locomotor patterns are not connected up and developed, certain socializing patterns do not fully emerge. The same researchers then found that encouraging the stutterer to creep and crawl once again served to call up the immature motor patterns in order that he might develop them. As he did so, his speech and his other socializing activities improved.

The body teaches itself. That which is becoming conscious teaches that which is conscious, and that which is conscious teaches that which is becoming conscious. As our bodies move toward containing the excitement of our life experience, we refine our excitation in the process of transmitting it from one level to another. The learning process is our formative process. We open our boundaries and extend ourselves; then we integrate our experiences. First we become more. Then we know more and we express more; we live more.

A few years ago I gave up smoking. Smoking had been nearly a lifelong ritual for me, and during the days that followed I felt uncomfortable. And then suddenly I realized that I was having sensations in my throat and mouth that I'd never experienced as an adult. Stopping the smoking had re-eroticized my respiratory tract, and the experience of the new sensations made me anxious. When I recognized that that was what it was, it put me into the more-ness of what was going on as me, and this gave me pleasure. My discomfort was related to having sensations that I didn't recognize and couldn't fully appreciate.

To describe learning in terms of role-identification or problem-solving gives a very limited and misleading picture of what goes on in the actual course of one's learning process. To learn is to experience new patterns of excitation and to embody them. An experience of excitement in the chest travels downward and expresses itself as sexual feeling and pelvic reaching. It travels upward and expresses itself as words: I love you. The eyes see with this same vibrant quality, the arms reach out to the loved one, and we are changed.

IMAGERY AND SELF-SHAPING

All too often, our way of being somebody is to be somebody else's body. Our way of being embodied is to be Daddy's ideal of a good girl, Mommy's ideal of a good boy. We learn to mimic people, especially people who are successful by society's standards. "Act like your father." Or, if father is a failure: "Don't act like your father."

And then we live a lot of fantasy, a lot of if-only. "If only the world were a different place I could be happier, more outgoing, more sensually alive." Or we escape into booze, the movies, compulsive eating. Instead of inhabiting our bodies, we both abuse and run away from our bodies. And things happen to us. We become victims.

Fantasizing and mimicking and taking assigned roles are methods for avoiding self-forming. We practice these methods until we identify with them. Most people who come to work with me have experienced very little other than self-avoidance. For various reasons they have been unwilling to disengage from these static attitudes, stereotypes of performance which announce, "This is the way I must be."

So I invite people to get at the source of themselves. I ask them to evoke their own excitement, call forth their own sensations. I encourage them to express themselves physically and to feel the emotions associated with their expression. I encourage them to move parts of themselves that are rigid, and to recognize the thoughts related to their stiff body attitudes. I invite them to uncover and, above all, to *live* their own excitatory patterns and rhythms. Out of their living grows their understanding. All of this becomes possible as they challenge their assumptions, their roles, the ways they do things.

When we are willing to experience ourselves, we don't live anybody else's life. Nor do we lose ourselves in fantasy. As our bodies become freer, less constricted, we begin to shape ourselves imaginatively rather than mimicking and fantasizing. Who developed Babe Ruth's batting style for him? Who inspired Columbus to set sail? Who teaches lovers how to make love?

To Know or to Grow

To form is to grow. A crystal expands itself in an additive way; a tree, in a geometric extendedness. We humans grow by increasing our motility and coordination and by inventing new behavior, new shapes and feelings and responses. The intricate connectedness of developing shape and responsiveness that we usually call growing, I call forming.

Growing is more than knowing. It is more than the

collecting of data and concepts. Our emphasis on knowing causes our brains and upper circulatory systems to grow while the rest of us remains infantile. It is our emphasis on knowing that leads to our forming so many bodies that are out-of-shape and misshapen. Our focus on the brain gives rise to beautiful heads, to well-formed thoughts, and to misused bodies with a narrow range of feeling. By living cerebrally, devoting our energies to collecting data and abstracting experience, we image our narrowed feelings in the form of an impersonal and dispassionate body of knowledge. We picture the world as a feelingless machine, operating by a system of laws and rules, rather than as an alive, growing, self-shaping universe.

It's our emphasis on knowing that enables the brain to feel that it "has" a body. The cognitive separation of brain and body is created by the brain's capacity to bifurcate, to make distinctions, to categorize, and give value. In this way our brain can come to identify our aliveness with itself and to consider the body as a thing. Then, rather than *being* some-body, we *have* a body.

We want to know for the sake of power, so that we won't be victims to nature or circumstances in our inside and outside worlds. When we know, we are able to predict, manipulate, and repeat. But this is performing, not experiencing or expression. It's progress, not process. Its aim is power, not pleasure; control, not cooperation; domination, not satisfaction. To form, to grow, demands an emotional commitment. It demands a continually maturing expressiveness, a life-style that insists upon the pleasurableness of living rather than power.

If you want to know yourself, slow down. Stop what you are doing. But if you want to grow, if you want to form yourself, you must actively express yourself.

The choice between knowing and forming is a choice that many of us are making today, often without realizing that we are making it. We wish to know ourselves, so we accept disciplines that ask us to curtail our activity. By slowing down and stopping, we can come to know ourselves by abstracting from our experience. But we also cease to form ourselves. To form ourselves, to grow, requires that we be expressive, that we try to shape our situations. To form demands that we accept the risk of the unknown.

There's a difference between knowing who we are and forming who we are. The psychoanalytical movement has collapsed on the mythology of "Know thyself." Ever since the days of Socrates, to know oneself has been the foremost goal and highest achievement of our culture—but damn it, whatever you do, don't be yourself. Don't form yourself. Let us form you.

To grow is to change the shape of your living. It's never too late to grow, to fill yourself with your own life. I have two friends over seventy-five whose whole existence has been the continued shaping and re-shaping of their lives. One of them is now rapidly losing his eyesight. When I asked him how he felt about this, he answered with excitement as well as melancholy, "Well, I'm going to have to make a different life for myself." My other friend just divorced his wife. He told me he had been making himself sick by trying to compromise his urge to be solitary. So he had decided to stop compressing himself. He went ahead with the separation and let his boundaries open. As he told me recently, he is in the process of forming a new self.

Some people are so busy knowing, finding out, accumulating information, being introspective, that they are perpetual students and slaves. They have confused knowing and understanding, knowledge and experience.

The highest level of excitement occurs during self-expression, and this does not necessitate a diminution of self-perception. When you express yourself you generate the energy for self-perceiving. Activity alone does not do this, however; self-experiencing is the necessary ingredient.

You don't have to know. You don't have to search out an answer or find a way. That's the mistake most of us make: we are looking for knowledge for a ready-made way, instead of letting our feeling and our self-expression form our way.

The Shape of Our Own Way

Anyone who maintains a particular style of living has a stake in perpetuating a particular structure, a particular pattern of form and movement, a certain fixed expression of being alive. But if one feels one's own liveliness, then one is willing to leave behind what one has already finished forming.

Consider what happens in the act of making love. Two people come into juxtaposition, and in the sharing of feeling and movement, all that is currently expressive of these two persons is exchanged. If they are willing to abandon what should be there for what *is* there, the out-of-the-ordinary is evoked. Every encounter is a beautiful encounter and a satisfying encounter—unless they are looking to repeat a pattern, to re-establish a feeling that they recognize as "love" or "sex." Or unless they want to believe somebody else's criteria and chase somebody else's goals.

To be a human animal is to express one's unique aliveness with one's flesh and blood. Our nervous system is capable of withholding expression of ourselves, heightening our vibratory state by permitting our excitation to be released and committed selectively. To say, "I surrender myself to anything that comes along" is to commit myself lightly. It shows contempt for the part of me that expresses me. By being contemptuous of it, I lose it.

By releasing indiscriminately, I seek to rid myself of something unpleasant. I make no finite commitment; I simply give up the energy of my self-forming, and this is an expression of self-contempt and fear. When a man and a woman are making love, they direct their expression toward each other to build their pleasure. They participate with *each other*. They say, "Let *us* form our lives."

Our Children

Our children are the expression of our being alive and making love. And yet they are more than the expression of us. They are the expression of themselves. Children naturally form themselves around the expressing of their own feelings. They educate themselves, organize themselves, regulate themselves, learning from the moment they are born.

When we encourage our children to express their feelings—their laughing, their crying, their curiosity, their tenderness, their anger—they develop flexible, graceful, expressive bodies, and they learn that the world is a supportive place in which to live and grow independently. By direct experience, direct contact, they learn to take joy in the individual forming of their excitement.

The trouble arises when, nursing the presumption that we're the only ones who know what life is all about and how it should be formed, we try to inculcate our children with *our* patterns of expression. Those of us who are older are habitually telling the young how to live, which is a roundabout way of telling them not to threaten us by living differently. "Don't be alive in any way we don't approve of." We act toward the young as if their aliveness and their expressiveness were intrinsically immature, as if they were incapable of learning from their own experience.

In the name of Knowledge we dampen and channel aliveness. Our current system of education creates spasms. We cramp our children's bodies so that we can form their minds. The school system institutes a social contract between the kids and the teachers, and between the kids and adult authorities in general. And the contract is a contraction model. Learning becomes painful. Learning becomes a chore that requires discipline.

A child who is brought up under a restrictive social contract learns that in those moments when he is most intensely alive, in those moments when he is most expressively himself, he is in danger. He may be able to express his ideas, to speak his mind. But what if he walks down the street feeling so alive that he wants to reach out to people, to hug them and caress them? What happens if he starts touching them?

Our cultural contract teaches us to refrain from touching. It teaches us to identify with a restrained role. Even under these conditions, children retain as much aliveness as they can, but they become increasingly aware that it's a dangerous world for an expressive animal. An expressive animal threatens society.

And yet why should anyone accept the idea that he is a social threat, simply because he's alive and strong? In my work, and in dealing with people day to day, I used to pull back so as never to pose a threat. But by pulling back I conflicted with my own expressive excitement. So I don't pull back any more. If I threaten people, I threaten them. If they walk away from me, that's okay. What else am I going to do? I am a man before I am a healer.

• • •

Every day, I live what is present in my life. I participate with what is. I also die every day. I let die what I don't need—and this becomes my past. Nature displays this simultaneous living and dying and points up the paradox of continuous discontinuousness, of forming and unforming.

Expression is self-liberation. When I express myself, I am actively participating with others, connecting with my environment. And this is my pleasure.

I forge my excitement on the anvils of expression and experience. Experience and expression form my self, just as I form my world.

EXPRESSION

To live is to express one's self. Our expressive self is our formative self. Expression impels us toward more self-forming, toward experiencing more deeply. And yet, in the act of expressing ourselves, we are not aware of ourselves in the conscious sense. This kind of awareness comes *afterward*. Self-awareness arises during the self-collecting, self-containing, and self-intensifying that follows a significant expression of being alive.

Expression is the outcome of our enlarged excitation, the out-pressing of what we have taken in and contained. The excitement of our humanness continually attempts to satisfy our hungers by pressing outward in pulsatory waves—filling us with the environment and filling the environment with us. Our greatest excitement manifests itself wherever we have the least boundary to restrict it. And our boundaries are least restrictive at the edges of our growing expression.

To understand is to be intimate with. This understanding is born of expressing. To confuse it with the collecting of data is to miss out on our living selves. In the beginning there was living, not knowledge. Knowledge is reaped, abstracted after the event. It is not the growing of the fruit.

If I am forever aware of myself, I constrain myself. I hold my excitement back from expression, from self-involvement, keeping it in the stage of containment which encapsulates and intensifies my experience of past involvements. To base my identity on awareness is to diminish my desires and to constantly focus my excitation into an objectified, stereotyped present—thereby cutting myself off from my future, whose source is my ongoing hunger to experience my self and others. In expressing myself, I diminish commitment to self-knowing. But I do not eliminate commitments to self-experiencing. I invest myself in my formativeness, willing to reshape my notions about myself rather than clinging to a preconceived plan.

There is nothing wrong with knowledge. What's wrong is an incessant, unrelenting need to know, which is related to a compulsive need for power. In our Western culture we stifle desire—the hunger for experiencing. Instead, we push programs that teach us to know—and to want to know. That's our biggest stumbling block. By saying, "I must know," by structuring our identity upon what we do or do not prove, we build a stone wall around ourselves. We say in effect, "I can't *be* until I have the facts." This undermines the emotional needs of living.

Knowing comes from experiencing. Knowing as a goal divorced from living, or touted as a way of living, is a poor compensation for experiencing. Yet many people do not wish to shape their own lives. They prefer to be given the knowledge of how to live.

The Language of Expression

The formative process has three phases: expansion, containment, and expression. When I work with a person, what we do encompasses these three phases. First, I ask the person somehow to expand the range of his movements. Then, so that he won't be carried away by the unfamiliar activity, I ask him to slow down, to contain and savor his emerging shape. Finally, I ask him to express himself again, this time integrating his new feelings and perceptions into his actions.

The expansive phase of a person's forming is, for the most part, a pre-cognitive, pre-personal phase. Somebody involved in an expansive activity is not usually conscious of his participation. It is possible for him to be conscious of it, but more commonly his cognition does not begin to enter in until he reaches the limits of his expansion, or bounces up against something not himself.

At the point of limitation, when the person begins to inhibit and collect himself, he begins to sense his boundaries. He begins to know increasingly who he is. He begins to discover that "this is me, and that's not me." And his response to this discovery is either fear or added excitement; either he withdraws in order to contain less or he seeks to contain as much as he can.

Containment is both automatic and self-initiated. Think of the heart: when it has sufficiently filled with blood, it contracts. Think of the stomach: when you have had enough to eat, you stop eating. When you've had enough fun running and playing, you stop and rest. When the two of us work together and your excitement has been expanding, you sense a reluctance at a certain point to go on with the expansive activity. You begin to feel who you are and who I am, and this feeling grows stronger. And then you *know* who you are, and you may decide that you want to share this by releasing your contained excitement so that it blossoms in the world.

From the womb of self-containment is born the language of self-expression: our *yes* and our *no*. We can choose to leap into the air or to sit tight. We can choose to laugh or cry or sing or keep silent. Whatever we do to express ourselves, we are saying, at least implicitly, "I feel myself so strongly that I'm going to take a chance. I'll risk a *yes* or a *no* and see what comes of it." So we say *yes*. Or we say *no*. And we interact with the social world —encouraging the tested and the untested, the familiar and the unfamiliar to greet us as we shape ourselves.

When we express excitation that we have contained and intensified, our relationships with others begin to change. Others react to the quality of our expression in responsive ways. If it scares them, they shun it. Or they consider it from a distance. Or, if they resonate with it, they move toward it and it makes the connection deeper and more vital between the two of us.

During a group session, one of the participants was standing for the first time without having to constrict himself in order to hold himself up. He was experiencing an enormous flow of sensation and feeling, and it was contagious. A young woman sitting nearby said that she could feel the waves of his excitation wash against her and sweep through her body. As this continued to happen, she began to know him; she began to contact him, and then to connect with him. And then she began to know herself in the same connected fashion. It was a revelation to her, a communication of abundant excitement from without and within.

In and Out: The Dialogue of Breath

I work a great deal with people's breathing. Often I ask people just to lie on the bed and breathe. I watch them as they breathe in and out, in and out, and I begin to discern a pulsatory wave that has a unique emotional charge. No two breaths are ever the same. Never. They may have similarities, but they are not the same. Every connection that someone makes with the environment by inhaling and exhaling is different from the connection that went before it.

Breathing is a marvelous act. It's the bridge between two worlds. It spans the borderline between control and no control, between the taught and the untaught. In the civilized world we breathe more restrainedly. In the world of nature we breathe more spontaneously.

There is a very evident relationship between breathing patterns and individuality. Shortly after the arrival of our baby daughter, her breathing turned her pink, and we could actually see her birth as an individual. She was an impersonal creature until she took her first full breath. And the same is true for all people. An individual who will not fully inhale will not fully inspire himself, accept into himself the influx of his surround. An individual who inhibits exhaling will not fully commit himself, give himself trustingly to his surround. An individual who will not fully breathe restricts his individuality.

Chest breathing reaches into space. Belly breathing

reaches toward the earth. And when I work with somebody, I focus on the particular area where his breathing is restricted. During infancy, most breathing takes place in the abdomen. As breathing expands into the chest, the child becomes increasingly assertive. As breathing expands into the pelvis, the child becomes more self-assured and sexual. People who are afraid to cry or shout, or who are afraid of the feeling of "I," inhibit their breathing in their chests. Those afraid of sex or given to self-worry inhibit it in the pelvis.

We are born breathers who form our own ways of breathing. A person who acts submissively diminishes his range of breathing. A hysterical person overextends his breathing to the point of panting. The long-distance runner deepens his breathing.

Whenever it's feasible, I try to coordinate people's breathing with other forms of expressive activity. If a person has deep contractions in his chest, I try to help him open his breathing into this area so that gestures and vocalizings of assertiveness and self-esteem can emerge without being mechanical. If a person has constrictions in his pelvis, breathing into that area releases feelings of sensuality and groundedness which he can then live.

To Cry

Breathing facilitates crying, and people cry a lot in this work. They cry not only for the sake of catharsis; they cry for joy, from having made connection with more of themselves.

Crying expresses the full range of human emotion

from anguish to ecstasy. Even laughter is a derivative of crying. If you watch someone cry, in pain or for joy, you see that the whole person is convulsing rhythmically. It's interesting, because crying, unlike other forms of expression, almost always brings about the basic involuntary, pulsatory movements. Anger generally does not. A person capable of expressing anger is not necessarily willing to cry. It's the person unwilling to cry who is unwilling to practice his own freedom. It's the person who won't cry who does not fully share.

Anyone who has been around children has noticed that their every emotion is expressed by some form of outcry. Their cry is the voice of their bodily freshness. And in this respect adults are no different from children. As we become able to cry, our bodies become more capable of expressing our selves. As we learn once again to cry, we grow more and more willing and able to make joyful love.

We enter the world with a cry, and from that time on, our crying or our not-crying is part and parcel of our forming. He who never cries out is never heard. The warrior's roar, the lover's shout, the victim's scream evoke human response—and are heard by the gods as well.

The cry is the mother of all emotional expression: howls of anger, moans of sadness, sighs of tenderness, bellows of hunger, shouts of joy. We who do not cry ensure that our rigidities never soften, that we never become impressionable enough to form again.

URGES

Natural desires—urges to reach out, to play, to touch, to suck, to love, to taste, to cry and scream—fill us with the energy that extends us into the world for satisfaction. People who maintain chronic attitudes, unchanging patterns, narrow their feelings and limit their urges toward pleasure. They won't permit anything to happen to them. Oftentimes they don't know this. Many people say that they're having certain feelings, and meanwhile their bodily responses may be demonstrating very clearly that their wishes are quite different from what they think they are. Nowadays we're more sophisticated about these things, but it still happens that somebody actually tells me, while tightening his jaw and clenching his fists, that he is perfectly calm.

We are always being impinged upon. Impulses arise in us; new desires greet us each day. Most of these stirrings are aborted; they are not strong enough to penetrate the density of our selves and attract our attention. They hardly ever break into our lives unless they're of heroic or crisis proportions. And so we miss untold opportunities to enrich our lives.

Each of us is a sea of needs, a continuum of urges which we tend to ritualize and render more or less impotent. And unless we find ways to interrupt our routines, we become robots. Interrupting shocks our equilibrium, separates us from our embeddedness, makes us stand out from the world of routinized behavior. The challenge to our boundaries calls forth responses, fresh impulses.

Impulses are continually poking at us like little spearheads: potential pleasure-bringers. But most of us need crises to jolt us out of our stabilized lives. And so it is necessary now and then that others interrupt our boundaries until we learn that our own desires, our own urges do this for us. They unform us and re-form us.

• • •

Tom was short, barrel-chested, and hairy, with a very taut torso. A prattler, he took outspoken pride in the fact that he was tough and realistic. He joined a group that was working on emotional expressiveness because he felt that he was insufficiently responsive to others.

Tom: . . . Yes, I'm aware of that. But I have a certain amount of feeling for my old lady, and the love that I get, I guess. I've also been in a totally new environment for the last year, and I feel that—

Stanley: Wait a minute. You've got a wise guy in you. And for the past twenty minutes that wise guy in you has been undermining every approach to you. Instead of yes-butting us again and again, canting your head, looking at us patiently out the corner of your eye, I wish you'd bring that distrustful side out front. You could simply say, "I don't believe that; I don't experience it."

Tom: Yeah, but you may be imposing your trip on me, which may not be valid at this time. I don't know that anything in this workshop can do anything for me. Anyway, I feel that you're playing a game. And I feel that you're in the head, Stanley.

Stanley: That's not what you feel. That's what you think.

Tom: I've been feeling some sadness, some anxiety.

Stanley: In regard to what?

Tom: In the sense of not being able to reach through to you.

Stanley: Could you express that physically? More dramatically?

Tom: I'll try. But I feel it's hard . . . to work with what I feel.

Stanley: Do you know what position your head is in now? The head tilted back and the eyes up? It's an attitude of pride, or scornful disgust—looking down at us as if you'd like to vomit.

Tom: You want me to vomit?

Stanley: Well, if that's what you want to do, do it.

Tom: [makes loud vomiting noises] . . . All right now?

Stanley: Oh, lie down and take your shirt off. Bring your arms to your sides and reach up, as if you were reaching to someone, and open your mouth and breathe. And make the sound "Aaaaaahhh!" Keep doing it.

Tom: Aaaahh! Aaaaaaahhh!!

Stanley: Can you feel any vibrations?

Tom: I feel it here in my throat and a little farther on down, like a longing. Aaaaaaahhh! Aaaaaaahhh! It's in my upper chest. Aaaaaaahh! It feels better in my throat than in my chest.

Stanley: Say, "Why don't you respond to me?"

Tom: Why don't you respond to me? To me!

Stanley: Where do you experience that? And how?

Tom: Here, in my chest . . . a pain. . . . Dammit, respond to me!

Stanley: Now say, "I'm disappointed" again.

Tom: I'm disappointed. I'm disappointed, disappointed. . . . My lips are quivering. . . . I want so much to get through to people and I never do.

Tom started to weep deeply as the quivering spread through his body. His mouth and arms reached out. Interrupting this man, interfering with his yes-but and wise-guy attitudes, stimulated him to experience more of himself. He let his feelings break in on him. He experienced his urge to extend himself. He was able to cry. Until then, his arrogance had protected him from his desire in such a way that he could not form himself to deepen his satisfactions.

REVELATIONS

Claude

Claude does not focus with his eyes. This diminishes the anxiety he feels in his eyes. He fears seeing and being seen. When I look at him directly, he feels himself shrink.

As he moved his hips, excitement started to build up in his legs and pelvis and manifested itself as itching. The itching provoked anger. His reaction was to kick. After a while his pelvis began to make involuntary motions, but he was pushing it as a rigid unit with the muscles of the lower back rather than undulating pleasurably with hips, legs, and feet. He seemed to be trying to get away from his excitement by compressing himself. When I suggested that he stand up and rock his pelvis rhythmically, that he reach with it and then regather himself, he shoved it forward and yanked it back. This, he thought, indicated how strong he was. In actuality it revealed his fearfulness, his extreme lack of confidence.

He knots his diaphragm. He draws back and locks his mouth and throat to reduce excitation which makes him anxious. When I ask him to express his anger, to shout and scream, his most common response is "I can't, I can't. I'm afraid I might lose control. I might kill somebody, or I might get done in as a result." So he squeezes himself, and then he reads back the squeezing as a feeling of impotence.

He holds his energy tightly in his head. When he releases it, he gets dizzy and has sinking feelings which flood him with a sense of dying. So he cramps even tighter.

Whenever our talk becomes intimate, he stops breathing deeply and stiffens so that he won't have to feel himself. He braces himself, an attitude that expresses "I can take it! I'll do it!" But this merely covers up a much more profound feeling of "What's the use? I can't. I don't dare."

I could not help this man as much as I wanted to. At the time, I didn't sufficiently understand that without his squeezing, he'd collapse. I didn't realize how deep were his feelings of defeat. When he became assertive toward me, I did not give in enough to encourage the forming of his confidence. I reasoned with him, which reinforced his old passion for being submissive.

Sarah

Sarah complains of stomach aches, vaginal dryness, and pains in the lower back and neck. She has spasms in her stomach that make it appear as if she is trying to defecate. She says she feels frantic.

Her thighs are flabby on the surface, but deep inside, around the bones, I could feel that the muscles were very tight. I put pressure on the top part of her thigh muscles, which had the effect of deepening her breathing. She told me that she felt a slight vibration in her pelvis.

When I asked her to rock her pelvis, she could not. She arched her spine and locked it back. This arched-back attitude allowed her to breathe into her abdomen, which protruded loosely, but it also tightened her chest, shoulders, and neck. Her face grew hot and agitated. She was somewhat aware of the contractedness, but she didn't know what to do about the self-choking.

Today I asked Sarah to lean over backward and put her palms against the wall. This stretching began to release the contractions in her thighs, and she began to make slight pelvic movements forward with each outbreath. She then reported a warm flow in her legs and a vibratory sensation in her upper back, and she told me that she had learned a little bit about how to help her excitation move through her feet and into her legs. Her lower back was still cold, however.

It was today also that I grasped the relationship between someone's being excited and my being able to experience this person's excitation beyond the skin. Previously, Sarah's low level of excitement had made it difficult for me to perceive her excitatory field.

Today she let her breathing heighten, but her pelvis was not involved. She still cramps her thighs. When I massaged the back of her neck to help her let more sensation enter her head, her pelvis began to move—but very slowly and cautiously. She appears to be overexcited, yet stagnant.

She told me she didn't know what to do with her pelvis, it felt so uncoordinated. I said to her, "Why shouldn't it be? It's like a baby taking its first steps." It is significant that after having experienced this pelvic movement, she was able to express anger at me for what she said was irony in my voice.

As Sarah stopped grabbing in her thighs, she began to cough. She became aware of wanting to expel something from her chest, and I suggested that she scream. She did, and her jaw began to vibrate. I asked her if she wanted to bite, because that was how it looked. But she denied this.

She told me about having very little perception of her lower limbs. She said they felt cold and weak. I massaged her knees, calves, ankles, and feet, and she began to cry. It was a laughing sort of cry, cut off somewhere down in the throat. Her breathing intensified, and when she stood up she began to vibrate.

At this point she was experiencing intense pain in

the small of the back. The pain frightened her, but she accepted it and let go a bit. This very slight letting-go in the small of her back led to increased vibrating and even deeper breathing. As she remained standing, the vibrating began to extend into her pelvis, and she experienced distinct sensations of enjoyment and satisfaction throughout her pelvic area.

Suddenly she leaned forward and squatted. Her back went rigid. Her pelvis locked. Then she collapsed to the floor and wept softly. She told me that she cried because the expanding of her excitation made her feel pleasurable and young; and for her, to feel pleasurable and young generated feelings of helplessness. It was wonderful, she said, but she was terrified that her back might crack.

Today I placed my hands on her pelvis to help her localize the hip contractions. After some difficulty, she finally did so. Weak excitatory waves began to travel slowly and gently into her pelvis and thighs. Her breathing deepened somewhat, yet there was a definite constriction in the throat. When I touched her there, her breathing grew quite deep. It turned into a growling sound, and she made sucking motions with her lips. Then she put her hand on her genitals. Later she told me that deep breathing provokes sensations that tend to release the contractions in her throat, eyes, and brain, enabling her to feel pleasure.

Today, to a much greater extent than before, Sarah accepted feelings into her thighs and pelvic area. Her field was a brilliant green, clear and vibrant like the sea in sunshine. She could feel herself streaming, and she told me that she was sensually excited, although she felt that if she continued to expand she would suffer some sort of catastrophe. By an effort of will, she tried to prevent her excitement from entering her head. When she stood she was extremely anxious, breathing fast and heavily, shaking, struggling to get rid of her liveliness.

I understood that her increased excitement caused her anxiety because of a fear of disapproval, a fear that her body's sensations were not accepted by others. Sarah could not tolerate an increase in excitement when she experienced it in the form of a need for touching, an urge toward being sexual. She registered it as anxiety.

I placed my hands on her to affirm that she wasn't alone in this dimension of her bodily life. This encouraged her to feel that her sensations were natural, and she was able to receive them into the conscious areas of her brain.

Sarah seems to be starting to form herself. She seems to be growing into somebody.

CONTRACTION AND EXPANSION

Harry came to see me in 1959—severely contracted, caught in a tube-like existence in which there was little room for the satisfaction of expanding and growing. His life was an unbending line of frustration.

We began to work on his contracted bodily attitudes, which gave shape to both his inner and his outer existence—to how he felt and thought, and how he moved and behaved. As Harry learned to give up his contractedness he let himself expand, reinstituting the process of expansion and contraction which formed a new body, a new somebody.

• • •

Harry came to me with complaints of a draining osteomyelitic lesion in his leg, a condition that had existed for several years and could eventually have led to amputation. He described himself as a homosexual with a

Protestant, middle-class upbringing. At age thirty-nine, his main dissatisfaction in life was that he could not form a lasting love relationship with either a man or a woman.

Harry's basic expression was stiffness. His stiffness was his way of being, his way of doing, his way of feeling and experiencing. His bodily set was compressed and ramrod-straight. It manifested itself as proud defiance. However, as we later discovered, this pride concealed feelings of being helpless and easily manipulated. Fearing his weaknesses, Harry stiffened himself.

He was severely constricted, deeply compressed, as if there were a rigid tube from his mouth to his anus. He was built like a pipe with stuffing packed around it, a pipe that funneled everything from top to bottom, down the hatch and out again. And he depicted his life experiences in the same fashion. He'd feel some excitement, and the excitement would run down the pipe and be gone almost as soon as it had begun.

The ramrod musculature limited Harry's mobility. His ability to expand was impaired. A frustration-oriented person like Harry does not permit himself to expand. He funnels energy toward the world, but in a propulsive, diarrhetic way, not in an expansive way. The fulfillment-oriented person is not afraid to extend himself.

Harry resembled an aged foetus. His head was too large in proportion to his body, the forehead swollen as if the brain were over-activated. The mouth and jaws looked sensual but misused. There was a yoke around the neck, a collar of terrific constriction which separated the arms from the torso. The arms were spindly and disjointed—hands, wrists, and forearms able to participate in only a limited flow of integrated activity. Their expression was the grasping and snatching of a frustrated child. To avoid disappointment, Harry kept his contacts peripheral and at a minimum. He used his long arms not to reach further into the warmth of the world, but to grab from the world and to hold himself away.

His chest was hairless, depressively flat, and darker-complected than the rest of his body. He breathed primarily with his upper chest, but aside from that it revealed very little movement. It was here that Harry still felt the rejection he had experienced as an infant. The flattened chest was a perfect expression of his heartfelt despair. Bursts of anger, futility, and self-loathing emerged from this part of him as it loosened up.

The lower ribs were widely spaced, indicating a continual pulling-in of the stomach area. Harry told me that a bulging abdomen was feminine and disgusting. "You have to keep your guts held in," he said. Harry didn't feel much with his guts.

The top half of his body was contracted but overactive. By contrast, his buttocks and legs appeared soft, but the tissue showed little tone. The thick layers of toneless fat expressed a passiveness that belied the ring of tension around the root of his penis. Harry had high spastic arches, and he also kept his legs pressed together, pressing himself as far off the ground as he could squeeze. The osteomyelitis that would not heal was in part pointing out his unwillingness to come down to earth and allow excitement to flow into his legs.

The rigid, infantile structure of Harry's body set severe limits on what he could do and would do. His constricted attitudes narrowed the possibilities available to him. His tubular form was unable to contain pleasure, unable to experience pleasure on any level beyond that of the simplest, most superficial excitation. He perceived on the nerve level: he experienced nerve pleasure instead of gut pleasure. He had the pleasure of culture-pride, ambition, one-upmanship. But he did not have the pleasure of satisfying his deepest desires: for contact, for warmth, for fuller self-expression and achievement.

Harry felt that he was trapped, that life was hopeless and he was useless. But with his immense head-energy he rationalized his way of being. It was a philosophy of restricted flow. Terrified of his own expanding, Harry felt that he needed to be an immature child in order to survive safely. This powerful need to remain immature gave form to his body, as well as to his feelings and his life-style.

• • •

Making the connection between how Harry looked and how he expressed himself verbally and actively was a very exciting discovery for me.* I started to see that the tightness throughout his torso served as a ground for his compulsive behavior. His inside and outside had the same energetic qualities.

Harry began to tell me about his inner rigidities: his fear of being drowned, his fear of being overwhelmed and swallowed, his fear of losing his penis, his fear of being squeezed and suffocated and not having room to move. Everything was compulsive. His contact hunger was as compulsive and as intense as his self-hatred, his feelings of futility and nothingness. He would escape

* Wilhelm Reich reported this discovery in *Function of the Orgasm.*

into fantasies, or he'd spot some injustice and go into a hysterical rage, but his rage never changed the situation and he'd sink once again into despair. He said he cultivated tension in order to "feel alive." He experienced the excitement born of tension as aliveness; anything else left him with depression and a sense of impending death.

And yet he told me, "The minute I feel sexual tension I have an imperative urge to get rid of it, to get it out, to get it off me." He couldn't bear being alive too long. He couldn't stand the possibility of frustration or disappointment. So he would masturbate—right away—or he'd run out and suck somebody off, or he'd find somebody to suck him off. He even seduced the boys in the school where he was a teacher. No relationship lasted more than a couple of suck-offs; that was the way Harry structured his world. His existence centered around the mouth and sucking. He lived in a world of suckers.

He was a shoplifter too. He liked money and material objects, and if he wanted something in a store he took it, immediately. He was unable to contain himself. It all went down the pipe.

To compensate for his insecurity, Harry was very ambitious. He was interested in playing politics, in power and control. He took sides strongly and constantly fought with authorities, displaying his fundamental rebellion against the good-boy role. Being a good boy meant pleasing others and satisfying their expectations. Being a good boy meant not reaching out, not asking. It meant clamping down on his excitation and dumping it. It meant snatching what he wanted. Harry deeply resented having to do this, and yet the sustained tenderness of reaching out was something that scared him even more deeply. His hands had become distancing organs.

Harry did express some tenderness, but the tenderness he expressed was outweighed by his compulsive hungers. His tender feelings turned into clinging. His reaching turned into a hungry, destructive snatching and grabbing and clawing. Even when he tried to be kind to people, he ended up pinching them and grabbing at them.

• • •

Harry's body revealed a number of non-conscious organismic choices, choices that were necessitated by and expressive of conditions he had met in early life. In order to survive those conditions, he had learned to confine his excitement to his head. Harry's entire expression was vertically oriented, directed toward a proud cerebral uprightness that was formed by his tube around his spine. He fled from his bodily fears into concepts, idealizations. His life took place in his fantasies and thoughts. And his sexuality was in his head, as pictures that he thought would bring him satisfaction.

It was clear to Harry that he had to keep his excitement locked within the walls of his cranium. The experience of inhabiting the whole of his body was terrifying for him. He was disembodied, and he dreaded letting himself be bodily. He would not permit his excitement to move down out of his head and grow toward the earth.

Harry's body form related closely to his interest in interior decorating. He based his sensitivity on an attitude of pride and a great fear of expandedness. And from this bodily state he created his art. It was overdelicate—a finely exquisite art. There was no womanly force in it, none of the round, strong, forceful passion of earth and mother. His art was fragile. It was thin in form, delicate of line, holding forth a less fearsome but brutally restricted world.

• • •

Harry was scared of loving and being loved. His ejaculations were as fragile as his art—highly excited on the surface, but lacking the feeling of full-bodied connectedness. His chronic rigidity cut off his pulsations and streamings, leaving him with only his vibrations. His consciousness of feeling was limited to his senses, to his pictures, to the periphery of his being.

Love to Harry meant slavery and dependence. He had sacrificed his bodily self to his family's insistence that he relate to them as a non-sexual person. He had surrendered his masculinity and his freedom by assuming the severely restrictive role of "nice boy." And he was frightened of having to make more sacrifices, of having again to become dependent, pleasing, contracted.

To love and to be loved were both dangers to his existence. When somebody got close to him he felt that he was going to be caught, trapped, enslaved. He wanted to be loved episodically. He wanted hotel-room sex, with none of the ongoingness of a long-term relationship. He preferred to play gas-station attendant: I service you; you service me. He wanted no demands placed on him. He did not wish to love another, and he certainly could not love himself.

And yet the chief paradox of Harry's life was that everything he did was a desperate search for love, a desperate search for the feeling of being loved. He yearned for the *feeling* of being accepted. He longed for the love of someone who would give him security and support. He wanted to feel warm, he wanted to feel relieved of his hurts and his tensions, he wanted to feel that someone received something from him and gave him something in return. He wanted to nourish himself on these feelings.

He pretended to do this for others. He taught school, and he would say, "I'm taking care of the boys. The boys need a father, and I'm giving them what I never had myself." But actually—and Harry later told me as much—it was the boys who were caring for him. By acting concerned, *he* became the loved one.

Harry was the one who needed the love and acceptance. Yet he had a terrible time admitting this to himself. He'd go out and get sucked off, and then say, "See, I'm giving something that's accepted." The pick-up sexuality expressed Harry in many ways. But mostly it expressed his desperate way of getting and giving love.

• • •

Harry's behavior was an attempt to have pleasure like a child. He had never been satisfied, never had the experience of being filled. All he recalled was the stress of deprivation.

Pleasure plays a very important part in the process of how one forms one's self. We do not do that which is painful unless we absolutely have to. Basically, our pleasure is tied in with our growth and development. But if our formative process is drastically inhibited, our capacity for pleasure is inhibited just as drastically. And if our activities become associated with frustrational feelings, it takes special work to reorganize ourselves around pleasure.

Harry had a great deal of difficulty in permitting his reorganizing, the re-forming of his body, and there was an analogue to that in the great difficulty he had in permitting pleasure in the act of love. His pleasure was confined to getting rid of his excitement, to making quick contacts and getting served.

All the same, what gave Harry the willingness to change was that he began to feel that his own constrictions were strangling him. His efforts at satisfaction were not giving him pleasure, and yet he could not expand into another way of living. In entering into the work we did, he began to discover the possibility of being joyful. He experienced that containing gave him pleasure; rather than compulsively trying to get rid of his excitement, he found that the sustained contact of staying with himself or with others gave him a more deeply satisfying shape. In this fashion he moved toward taking responsibility for his own life feelings.

• • •

My main concerns with Harry were the yoke around his neck, the tube-like contractedness that ran the length of his torso, and the interference to the excitatory flow throughout his pelvic-genital region.

I began working with the pelvis, helping Harry to recognize the passivity there. We soon found underlying spasms, including the tight ring around the root of the penis. The squeezed legs and the circle of tension prevented the movement of excited feelings into his lower body. They also prevented his perception of these feelings.

As we worked to open the squeezing attitude, something happened, something visibly changed. There was a downward flow of excitement, a flow of blood. I asked Harry to tell me what was going on in him. He reported the feelings of aliveness, but he was also experiencing acute anxiety. He didn't know how to deal with these new feelings, and they were tremendously upsetting. He could not maintain an awareness of his own aliveness. Either he drifted into fantasies—fantasies of fleeing across an open space, fantasies of devouring and being devoured—or else he passed out to get away from sensing how heavy and dead the rest of him felt. This was precisely how, as a child, he had avoided feeling rejected and terrified.

Awareness is directly related to mobility and motility. Harry's body revealed some areas, like his mouth, that were more mobile. These areas contained his excitement; they had good tone and motility. And these were his areas of consciousness, his self-expressive areas. In the other areas of his body there was fear, and therefore consciousness was impermissible. Harry's unconscious areas, like his pelvis and neck, were hardened and dull.

Where one has deep muscular contractions, one's body image is distorted. If people draw pictures of themselves or imagine themselves or scan their bodies, they skip over the parts of themselves that are deeply contracted. They are numb to these parts. These parts are disassociated from their bodily perceptions. Because of the chronic contraction around the root of his penis, Harry didn't feel that it belonged to him. He told me again and again that he wished it didn't belong to him, that he wished he didn't have a penis at all.

Whole areas of movement and feeling, of response and decision, were outside Harry's scope. He did not know they existed. For instance, he didn't know that he could extend his body in any way other than straight up and down. He had no idea that he could rotate his shoulders and look behind him while his pelvis was still going forward. If he turned, he turned on a pivot. The flexing and extending movements involved in twisting himself around were eliminated from his repertoire because his body image did not sanction them. And so he contracted the choices that he was able to make.

• • •

As Harry began to let his excitement soften his stiffness, he also became conscious of the fear that was causing him to stiffen. But it took him quite a while to develop a new body image that would enable him to maintain the aliveness without succumbing to the fear.

A flow of energy into his chest produced a cold awareness of the love that he lacked, a fit of crying, and then rage or depression, or both. A flow of energy into his pelvis produced disgust at himself and his sex, followed by intense anxiety. Harry found himself afraid to get an erection and afraid not to get one. In his moments of helplessness he declared how much he wanted to suck, to grab, to yank in. Sometimes he blacked out, agitating his body hysterically. At other times he threatened to attack me, and this led in turn to his wishing to damage himself, the self that he hated.

At one point he recounted an amazing story. I don't remember how it came up, but this is what he told me: "At age seventeen I went out and joined the Navy. I came home and broke the news to my mother, and that night we lay in bed together. She lay next to me, her ass to my front. And I got so excited, so charged, that I almost went berserk. The only thing I could do was deaden myself." I asked him how he had deadened himself, and he said, "I stiffened up inside, all over." He dared not have an erection, so he tightened up and formed the constriction at the root of his penis that kept it out of touch. He made himself rigid and blank.

At this juncture we both understood that Harry could permit excitation, but he would not permit the excitement to develop into feeling and action. We both gained the insight that a person can have a hell of a lot of excitation and very little feeling. There's a slipped stage in the formative sequence. Most of the containment stage is missed, and the person goes straight from excitatory expansion to reflex, unparticipating discharge —a knee-jerk response to stimulus. In Harry's own words, "When I feel excited, I gotta get rid of it."

The women in Harry's family lavished a great deal of exciting attention upon him—teasing him, titillating him. But if Harry let that excitement build up, the women rejected him. So he learned to be afraid of it, and in order to be safe, he learned to get rid of it. He formed the pipe, which contained nothing. Maintaining the pipe got rid of what threatened him. But then Harry *felt* dead. He could feel so depressed and un-alive that he would panic; he would rush out and find someone to suck, to charge himself back up again.

We really went through the mill together. But after

several months Harry began to perceive that under all his hysteria was a longing and a sadness and an emptiness. Gradually he allowed himself to experience these feelings, and to experience the qualitative difference between excitement and feeling. As he softened enough to feel his emptiness, he began to recognize how emotionally impoverished and undernourished he actually was.

• • •

Another story he told me: His mother used to get undressed and go to the washbowl to wash. Harry loved to see her naked there; he used to masturbate while watching her. One time, wanting to approach her more intimately, he set himself up so that she would catch him masturbating. He lay on the couch directly in front of the bathroom door. His mother came out of the bathroom, saw him jerking off, and didn't say a word. She walked on by him as if nothing were happening. He never forgave her for that.

Harry felt deeply and personally cheated by the women in his family. As a baby, he had been fed with an eyedropper. His mother had been imposing, demanding; she continually threatened to withhold the love that she was never really able to give. But it wasn't only his mother. His aunts, too, during his growing years, consistently maligned the men in their lives. His father had run away and his uncles had gone off to the World War, and the women liked to sit around the house together and disparage the men. They made all forms of masculinity seem unacceptable. Harry recalled, for example, that they dressed him in girl's clothing. And he resented that terribly. But even stronger than his resentment was his fear of being rejected, so he learned to tailor his movements to girlish patterns that the women would find acceptable. He also learned to think the way they did. He saw all men as weak slaves incapable of being true to themselves. By escaping the slavery of manhood, by taking on the sexual role of an effeminate child, he freed himself from the burden of being a husband and provider.

The work we did helped Harry to discover that his homosexual "freedom" was itself a slavery. He saw that he was enslaved to his need for episodic contact, that he was in bondage to his conflict between wanting to please women and resenting the obligation to please them. As we did exercises to help him experience his legs and pelvis, he began to express the feeling that loving meant penetrating and being loved meant being penetrated. Before, he said, he had avoided penetration, even in his encounters with men. He had preferred to stiffen his spine and suck.

Later on, Harry went on vacation with some women schoolteachers and initiated going to bed with one of them. But he had difficulty taking part in the lovemaking. He was still embodying a tremendous amount of resentment. He still felt that while he was entitled to love, the woman didn't deserve to get any love from him. He resented paying the bills; he resented giving up his sperm. Mainly he resented what he saw as having to perform. He could not let himself relax sufficiently to understand that making love with a woman was anything other than performing.

• • •

The movement I was encouraging in our work together was a flow downward, which involved asking Harry to release some of his contractions to permit the possibility of a new kind of fulfillment. As things were, Harry had only two ways of expressing the downward movement of his excitement. One way was out the pipe, and the other was the holding or freezing attitude. Both were unsatisfactory. So we worked on developing patterns that would allow him to experience for himself that for many feelings, the pelvis is a better container than the head. This meant that Harry had to be willing to let his old world die—the world that he had formed so vertically and rigidly—and at the same time he had to be willing to let his new world expand, willing to explore the world laterally as well as up and down.

Much to his surprise, Harry began letting go and accepting excitation into his pelvis. He could now perceive that his formative choices were conditioned by past events. He could also see that the immediate reality might be experienced without reference to past negative conditioning.

In the past Harry had rationalized his over-reactiveness by calling it "sensitivity." He saw himself as one of the elite. But when he opened the circle of tension around his penis and really began to feel that he was connected to a cock, he began to feel much less bound to the ideal of being forever gentle and sensitive. He felt the thrust of his pelvis and penis, and all of a sudden he was interested in penetrating. He was moved to express the assertive side of himself.

One day a great change came about. I asked Harry to assume a position of hyperextension, leaning over backward with his arms outstretched. Suddenly a contraction in his abdomen gave way and he started to scream. He couldn't breathe and he totally panicked, turned different colors, and finally collapsed to the floor. This wasn't one of his ordinary blackouts. Harry had really become a helpless baby, unable even to stand. I picked him up, carried him to the couch, and listened to all his hurts come bursting through—all his hunger for love, his forbidden infantile longing for close contact and acceptance. He let it pour out of him, and then spent the weekend in my house letting himself be taken care of.

This was the turning point for Harry. Thereafter, he could begin to construct a world that to some extent gratified the deep urge for love he had been denying.

He was progressively able to free himself from the enslavement that his mother represented—in terms of both his dependence upon her and his rebelliousness against her. He grew stronger, more confident, and less impulsive. He outgrew the need for the stealing and the sordid sexual adventures. Even the osteomyelitic bleeding eventually stopped. It stopped after Harry came back from taking a trip and realized that he had gone away to escape the excitement of his own forming.

• • •

With Harry, I began to perceive how it is that bodily excitation forms a person, how the excitatory flow determines the body's configuration and the amount of charge that the person can comfortably handle. I also began to see what happens when someone begins to reorganize the flow of his excitement.

In working with Harry I proved to my satisfaction that our chronic rigidities do indeed restrict the flow of our energy. When Harry let go of his contractions, he was flooded with a stream of excitation and experience. Everything he did to re-establish the circulation of his excitement made changes in his life—physically, psychologically, and socially.

For Harry, learning to accept his body and to work with himself was an act of love, a willingness to love in ways that were neither humiliating nor self-destructive. He never grew to love a woman sexually; his hurts and resentments were too big for that. He finally reached the point, though, where he gave up his kiddie nonsense and settled for an ongoing relationship with a man. He hadn't been able to do that before. And I really understood: heterosexuality is not the only goal. Harry expanded into a relationship that had continuity and meaning. He was living a fuller, more formative life—accepting nourishment from the love of others, and expressing his own love so as to form a continuing connectedness for himself that gave him deep satisfaction.

• • •

When Harry first came to see me, he was someone of neuter gender. He had been frozen as a young child. In the process of watching him unfreeze, I had the opportunity to observe how a person begins to relate to the opposite sex, how a man's energy starts moving toward a woman and what happens when it does.

As a result of working with Harry I began to understand one other thing: a great many men hold to a contradictory attitude that simultaneously idealizes and abuses women. A man is taught to adulate his mother, to set each lady of his acquaintance on a pedestal and nobly protect her. And then to ill-treat her: to belittle her womanly strength, to define a subservient role for her, to keep her around for the purpose of pornographically fucking her.

Implementing the roles of what a woman should be goes along with pursuing certain roles of what a man should be. The majority of the roles presented in the fashion magazines, on stage, and in the movies are both idealistic and contemptuous. A woman is either a whore or a queen, a sex goddess or a skinny girl with short hair and no breasts. A man is a macho or a faggot, a gangster or a milquetoast, a roué or a clown.

Women, of course, are not entirely the innocent victims. A number of them, enamored of the situation, help to perpetuate it—showing scorn for the man who isn't tough, yet praising the homosexual for being cute and smart and sensitive. The women, too, agree to settle for fashionable thrills in place of allowing the gradual building and emerging of their own energetic excitement.

The female and the male roles are going through a crisis of re-formation. As the John Wayne image begins to deflate, the suppressed feelings of men wanting to be closer and more tender with others begin to manifest themselves, and these feelings are at first denied and covered up. Or, if expressed, they are expressed not in individual terms but in terms of a stereotype that everyone already recognizes, such as the ass-slapping of the football player or the clever effeminacy of the stock homosexual. As men give up their macho stereotypes and become less rigid, their capacity for tenderness increases.

But many of them lack the individual maturity and flexibility to contain and express these tender feelings. They are unable to let their tenderness deepen. So either they reject it out of hand or they make do with the role of being superficially sensitive. Equally, the woman who rejects the role of goddess and servant seeks pleasure by imitating male aggressiveness.

During the past few years we have at last begun to develop some very different images of what it is to be a human being. Nowadays a man can be sensitive and he can be assertive. He can be either one and he can be both. And a woman can be both. A woman doesn't need to surrender her sensitivity in order to be able to assert herself.

EXPANDING OUR SELVES

When our body loses its feeling, we lose our connectedness to ourselves and to others. And then we search for something to believe in. If our body does not feel the glue of connectedness, we try to locate God, life, somewhere out there. Or else, projecting our own deadness, we say that God is dead, that life on earth is doomed.

The person who is the core of *his* life is the core of *all* life. Because he feels what his own living is, he experiences and has faith in his expressiveness rather than in somebody else's expressed belief.

One effect of constricting our selves, our pulsations, is to create a non-rhythmic sense. We form a sense of stillness and permanence, and it seems concrete and secure because we have deadened the rhythmicities of our pulsatory movements. We narrow our experience, trying to eliminate the unexpected by creating an unchanging form, a static person. But the apparent stability of this narrowed lifestyle is only an illusion.

Many of us bind things together by means of thoughts and memories, by means of concepts instead of feelings. If we perpetuate this brain process we form ideals which we strive to live up to. Ideals have to be pumped up with our energy or they don't work; so we act "as if," or we force ourselves to act. This performing and forcing creates constrictions in us that we don't even know we have. All we know is that our lives are painful or dull or scattered or thin—somehow unsatisfying.

Trying to form connections with your brain alone inhibits your feeling life, restricts the expanding that reshapes your self and your relationships. This is how systems of belief flourish. They try to tell you about other people's feelings and experiences, how other people formed themselves. And while they may sincerely hope that you will have the experience and form your life in the same way, they do not encourage your own expansive forming. Systems in general try to do away with the mysteries of living. They tell you what the mysteries are and how to deal with them. They do not make you the mystery, nor do they point to the source of your own forming.

When I began to challenge my assumptions, ideals, and beliefs, both cognitively and muscularly, I began to experience an expanded dimension to my life. I started to understand that my experience and my expression were my truth, that my way of forming my self was my life. Self-expanding is exciting, and grows more exciting as it thrusts me toward individuation.

There are no salvation trips. What's there to save? When we live our own lives we take possession of our innate gift of expansion, and we use it to shape feeling, pleasure, and satisfaction instead of ideals and beliefs. We spend our lifetime becoming somebody instead of maintaining an image.

• • •

In an appendix to *The Psychology of Invention in the Mathematical Field*, by Jacques Hadamard, Albert Einstein says that for him, the primary process of perceiving is muscular and visual, and that he then elaborates this by the search for and development of a suitable language. Albert Szent-Györgyi, the biologist, writes that "Life keeps life going"—that life builds on itself, staircase fashion. Energy expressed as activity increases interest and creates more energy and more expression, even for the compulsive type who keeps busy living off his reserves.

Szent-Györgyi also points out: "When you don't use a machine, it lasts; when you don't use a body, it breaks down." When one is sedentary, the heart atrophies. Joggers, as well as lovers, have more energy. How they use it may be related to a system of values, but energy they have to open their own boundaries, to increase their own excitement, to form an expandedness of living.

Someone took me to a psychic meeting where a man giving a lecture said that in a former life he had been an English minister, and that when he died it had taken him three days to find it out because he was more alive dead than he had been alive.

• • •

Paul, who sat in front of me with his shirt off as he had so many other times, was different this time. His shoulders were not raised high in the manner of the scared chap he usually was. Nor did he wear the perpetual smile of the charmer that sought to avoid any straight response. He was telling me about the trip he had just come back from, an experience with his friends and family that he said had given him a feeling of his roots. And his body looked it, with his lowered shoulders, his relaxed neck and smile. Even more important, his feeling of having roots expressed itself in his lowered breathing. He no longer kept his breathing up in that compressed chest. It heaved with emotion as he told me that now he had a new relationship with his father, that he had discovered a strength in his father which transmitted itself to him. Paul glowed with this strong feeling of himself.

Then he went on to say that he didn't feel quite yet that he was back in California, that he wasn't feeling present at his job. He said he felt somewhat disorganized, and that it would take him a bit longer to start being his old efficient self again.

Gee, I said, here you tell me about finding a new relationship to your father, a depth of excitement and a deepening of your life's connectedness, and now you tell me you have to get used to being back! How could you not feel uncomfortable, unbounded, unnatural, different, strange? You opened your old self. You're becoming a different Paul. You have the chance to test the world in a different way, to completely renovate how you relate to the boss, to your girl, to me. So go ahead and form a new shape.

Loving

The expanded experience is a state of aliveness in which the heart overrules the head. We call this love. We can have a taste of this any time we want by trusting our bodies and surrendering to the immensity of our lives. When we do, we understand that our analytical picture of reality is not itself reality.

When a man and a woman come together, what happens is divine. It is not a return to the womb; it is universal sharing. What is implicit in the act of loving makes possible the continuation of living.

The expression of one's feelings forms the love relationship. To love is to drop one's roles, to be present extendedly. We drop the familiar patterns, and we take a new shape. I have never been in love without feeling my body in a new way.

We limit our ability to lovingly expand if we are chronically tight-muscled. We limit the experience of our womanness or our manness, and we limit our experience of the opposite sex. We do not enhance, enrich our lives or our friends' lives. At the same time, expanding is not unlimited. It is a deepening of form and feeling.

Once when I was making love, the unexpected started in me. I started pulsing, and the waves of pulsation spread throughout me. And then suddenly, in the midst of all this pleasure, I was seized by a paralyzing surge of revulsion which shrieked, "I won't give in! And I hate you! You're trying to sweep away the 'me' that I've labored to establish as my identity. You want me to give in to you, and I won't do it!"

I understood that this voice was an expression of my social contract for remaining in the favor of the culture. My contracted social role did not want to yield to the pulsing of me. And I knew the tragedy of living without being alive. I felt in what ways the living of my tame role-player was diminished in comparison with the aliveness of my untamed me. The social me perceived that, while it could be alive, it could not create its own life. And it resented the untamed me whose connectedness with the field of all life enabled it both to be alive and to be vitally formative.

At the same time, I realize that to live strictly in the undifferentiated field of universality is as lopsided as to live exclusively in the world of roles. To be entirely without boundaries is to miss out on individual pleasures and joys as well as individual hurts and despairs. To be boundless is to avoid being bodily, being humanly grounded.

A person may have an expanded experience. But his experience will not root until he plants it and extends its reality by containing and grounding it, forming his self and his social reality. What greater mystery is there than the mystery of being incarnated, the mystery of being in the flesh? Our bodily lives bound and unbound, embody and disembody, form us-the-mystery as we experience the collage of living layers called us.

• • •

I see life not as a dualistic struggle between two poles but as a resonating field of continually reorganizing excitatory patterns. We all pulsate connectedly in an intense sea of creation. Some of us are in it in a cramp; some of us, in an expansive way. But we're all in the same sea of creation.

This ocean of continuity shapes my inner and outer space, my bounded and unbounded self. There is inner space when I close my I, and there is outer space when I open my I.

I remember one day in Germany when I was with myself intensely. It was a courageous morning, and I walked over and hugged a tree. I looked at the tree and saw it vibrating, saw it pulsating—and I'm not talking about shaking in the wind. I put my arms around the tree. I felt it sending waves up and down its length, communicating with its vibratory pattern. I understood that this vibratory pattern *was* the tree.

I felt the tree vibrating and I felt my own self vibrating. Then I took a few steps back and began to perceive the tree again, this time as a vibratory reality that oscillated between feeling the tree within me and experiencing it at a distance from me. I was struck by how the tree and I were engaged in a kind of dialogue that continually shifted the way I "saw" the tree, the way I experienced it—as being sometimes subjective and sometimes objective; sometimes in me, sometimes over there.

I walked away awed by what had happened. And from then on I found that, more and more, I was able to sense how my own vibrating and my own pulsating expanded my world. Whenever I could sense the relatedness, I felt myself engaged in an ongoing dialogue with the field of my surround. I also felt that I was crazy, that this was no proper way to be. After all, nobody had taught me that my world is mediated by the dialogue of our resonating fields. I'd always been taught to fix the world in one place. I was told that a tree is *there*, and one goes to the tree—not that the tree and I connect by means of an undercurrent of feelings.

The Streaming of the Spiritual Animal

Love, which runs deep, begins with the streamings of one's body. One's streamings fill one's self with so much feeling that the feeling transcends one's boundaries and expands one's connectedness with other persons. Anyone who has experienced this has experienced what it is to be oneself and to be at one with the world, to recognize that everything is in flux and yet to feel at ease, to live in the world in an uncognitive way and to trust it.

How do we get there? We don't have to get there; we're already there.

If, somewhere inside you, there is a twinge of excitement, a sparkle of excitement—even if it's only a tiny spot in your earlobe—let it expand. And participate in your process of self-forming.

Each packet of excitement is fragile. It's not all that powerful. It's persistent, it's strong, and yet it's fragile. It shrinks in a negative environment. Excitement has the ability to hibernate when circumstances are not appropriate for its expanding and forming.

Sometimes a woman comes to see me and says, "I must be frigid. I've been working at it, but I have no feeling when I make love." In reply I say, "Now, look: first of all give up the idea of working at it, and just tell me, or tell yourself, what you *are* experiencing." The woman might say, "I'm experiencing the anticipation of having more sensation." So I ask her, "Well, where do you locate the excitement of that anticipation?" She may say, "I feel it in my head, I feel it in my chest, I feel it does not go past my navel." Her first step is to discover what is going on. Then she can begin to express this. And then expansion starts.

What attitude are you meeting the world with right now? Are you meeting the world with being cautious and distancing? depressed and withdrawn? bombastic and outgoing? Identify the attitude with which you greet the world, and locate it somewhere in your body. Is it in the back of your neck? Is it in your eyes? Is it in your stomach or in your shoulders or in your knees? Wherever you locate it, however you identify it, notice how it shapes your thoughts and actions, your self and your responses. The accepting of your present form encourages your process of expansion.

• • •

I understand spirituality as a heightening of the excitatory processes of the human animal. The religious experience, divorced of its hokum, is the vivid experience. It is our vivid experiencing, and it is the vividness of what we experience. Its depth and its intensity correspond with how deep and intense our streamings are.

Our streamings are our biological field, the field of our embodied life. To experience our streamings is the spiritual experience. To live our streamings, to participate in the forming of our lives, is the great mystery and joy of existence.

The most important thing we can do for our selves is to trust the body which forms us as somebody.

MY VISION

I use the bodily metaphor as the monkey bars to swing between the faces of my existence. There is the aspect of my existence in which I see you and talk to you, in which objects are recognized according to a consensus. And then there is the aspect of my existence that I can't figure my way into analytically. I come to the place where I can't take my thinking any further. My insights find their limit. At this point my intuitions flash to the surface and the pulsating and feelings intensify. This excitatory soup of intuitions, pulsations, and feelings, where my form comes from, is the great universe itself.

The inside of my body is a world of felt potential, a world of non-separation, timeless and endless. It's vital and pleasurable. I immerse myself in it, and then I withdraw and reflect upon the experience of my immersion.

There is a doorway, an interface between the faces of my existence, the aspects of my life—the public life and the private life of my body. The best way I can describe this interface is: "Here I am! And there is the world." Whenever I discover a new dimension of living, I experience it as always having been. There is the world and here I am, and everything else is a memory.

We use metaphors so that we can share our experience. I use a bodily metaphor. The bodily metaphor maintains the aliveness of the public world and keeps the faith of my private existence.

My livingness, my formativeness, grows out of the living ocean that our cells swim in. The experience of this excitatory sea gives birth to my vision.

• • •

On a night in 1959, while lying awake in bed, I experienced an event that brought biology and psychology together for me, crystallizing my understanding of the evolutionary process that goes on as one's self.

I was lying on my back, stretching the tight places in my body and feeling them release. Without my even knowing it, this initiated a wave of pulsating up and down the front of me. I felt my brain open. And I had a vision of a snake, green and vivid, that seemed to coil out of my intestines and up into my head. The next night the same event occurred, except that this time when the snake wriggled into my head it appeared to enter a silver ring.

While I could see that the snake was moving in me, I also felt that I was myself the snake. I oscillated back and forth between these two perceptions: objective and subjective. There was the snake out there, and there was the sensation of flow within me. I had vision, and I had feeling. And the pendulum of my oscillating perception created an egg-shaped field which, paradoxically, extended infinitely in every direction.

Over the next several years I improved my connection with my snake. I began to sense the relationship between its flow and my process of being formed, my process of becoming more than I used to be. I began to accept an excitatory spiraling up and out, down and in—an expression of my excitement that was alternately bounded and unbounded, contained and uncontained. I felt that my snake was my self. Whether I experienced it at my body periphery or in the depths of me, I could feel that it penetrated me throughout. And as I began to live this feeling, I began to develop a quality of being a continually forming and enlarging field of excitation.

One day in the fall of 1964, as I was feeling the brightness of my streamings, I became alert to my snake's flowing intensely from my bowels up to the area of my sternum, both inside and outside. I could see and I could feel the stream of excitation growing me, forming me into more of a heart-person. A few nights later, a restless night, I experienced first a pain in the chest and then the flow piercing through my diaphragm. My image was that of a snake bursting through the center of an orange slice. The whole of the next day I was enveloped in a circulating warmth that came from me and nourished me. My world was vibrant and intense, and poems streamed out of me—forming me, informing me.

Shortly thereafter I had two more serpent experiences which expressed similar patterns of growing perception and participation. On one occasion I was startled to perceive a snake lying coiled in a field around my pelvis. Later that same month, I experienced an uncoiled serpent rising and trying to enter my head. My reaction in both instances was panic until I understood that wherever it might direct its expansiveness—head, chest, diaphragm, pelvis—the snake was me, and I was friendly. The snake was me and, at the same time, it was forming an extended me, an expanded me. It was my present life, and it was my promise of more life.

• • •

The snake is my personal vision of the formative process, as another event will make clear. On this particular occasion I experienced strong streamings going up both sides of my body, and then I saw not one but two snakes. Originating from a single tail in my pelvic basin, one snake coiled up my right side and the other up my left side. It was as if they were carrying on a dialogue, and they carried it on into my brain. There they paused, as if shaking hands. Then their heads merged.

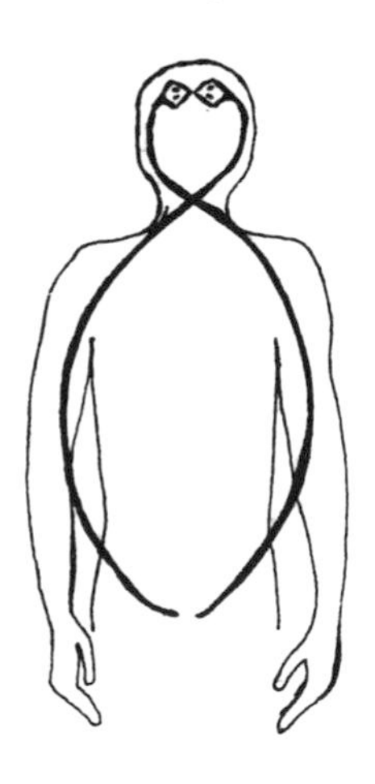

Looking at the pattern of the snakes' dialogue, I saw that it forms a loop of containment. The snakes open up from the pelvis and form a vessel. They cross over once. Then they close the vessel. Instead of going on to make a sequence of figure eights, they circulate back through themselves and through each other. An essential part of this circulation is the *pausing* of the snakes' heads before they fuse. It is in the pause that an agreement is mediated as to what will be unbounded, what re-bounded, and what newly bounded—what given up, what sustained, and what contained.

• • •

Though they are individual events, there is a relationship between the flow of the snake and the flow of breathing. Both bring about the feeling of an ongoing expanding and contracting. Oftentimes, by holding our breath we can more easily feel the pulsations and streamings that go on inside our body and sometimes on its surface. The movement travels from foot to head, head to foot.

As the snake and I formed our relationship, I began to feel that there was a qualitative and quantitative change which took place in the flow of my excitement from pelvis to head and from head to pelvis. The snake would coil up into my belly as an undefined impulse of love and thrust. Its impulse would soften as it passed through my heart, through the center of me. When it rose into my head, it would make a picture or an insight, a refined perception as distinct from a global perception. If I was funneling energy from my head on down, my thought-images would first unbound themselves; then the excitement of my snake would intensify in its downward course through my chest and into my pelvis, legs, and feet—illuminating and vivifying my connection with my undifferentiated surround, my creative void.

Gradually I learned that my snake is the continuum of my excitement, the pattern-maker of my imaging, feeling, and acting. It recapitulates, expresses, and re-expresses my molecular and cellular history, my past and present social living. And the circulating of my bio-history grounds me in the flesh. The snake is my oracle of the formative process.

• • •

The snake operates according to its own laws. In the snake world, the ordinary limits of time and space appear to expand. Space seems to be infinitely extended, and time is no longer confined to a linear progression. There does not seem to be a passage of time. Nor is there a quality of past-present-future. The flow of excitation reveals itself like flashes of a strobe light, or like view patterns in a kaleidoscope. There is perception and remembrance of events, but the events do not belong to the clock sequence.

My experience of the snake has given me to understand that there are different fields of existence. One part of me is formed in the world of time-space. Another part of me creates time-space.

What is personal is also universal. The snake that I perceive as my self in the process of forming is an archetypal image that almost all cultures have used to depict life's growing and transforming process. This evolutionary quality of expression unfolds as the waves of the sea, the double-helix of DNA, the twisting paths of blood vessels and nerves, the rhythms of waking and sleeping, standing up and lying down, feeling separate and feeling at one.

The snake is my image of moving between the horizontal world and the vertical world. It teaches me how I wake up and quiet down, how I make contact and disengage. It is an image of arousal and renewal that expresses the serpentine movement of my body's excitement weaving through the crevices of my brain. It's I talking to me about my own formative process, about my excitement becoming personal and then social. It resembles Henry Miller's image of walking across the Brooklyn Bridge, going from home to work, from home to away-from-home—from unconscious to conscious and back again.

Your own symbol may come to you in a dream, in a reverie, while you are making love. It recurs. It's your friend. And it reveals to you how you connect with the universe and with others. It is impersonal, but you make it personal. Miller's bridge doesn't belong to him, yet he made it his. The snake belongs to no one, yet I have made it mine.

APPENDIX

TWO WAYS TO FORM GROUNDEDNESS

At this point, if you're feeling a buildup of excitement from reading this book, let's see if we can ground it. Or if you're bored, perhaps you can create a feeling of charge. Or perhaps you would just like to experience what I have been talking about.

• • •

Stand with your feet parallel and about six inches apart, shoes off. Keeping your feet in place, bend your knees and spread them a bit farther apart so that you end up in a semi-squat.

Take your two fists, reach behind you, and push them into your lower back on either side of the spine. Keep pushing with your fists in the small of your back until you begin to get a sensation of release in your lower abdomen. Let your buttocks protrude.

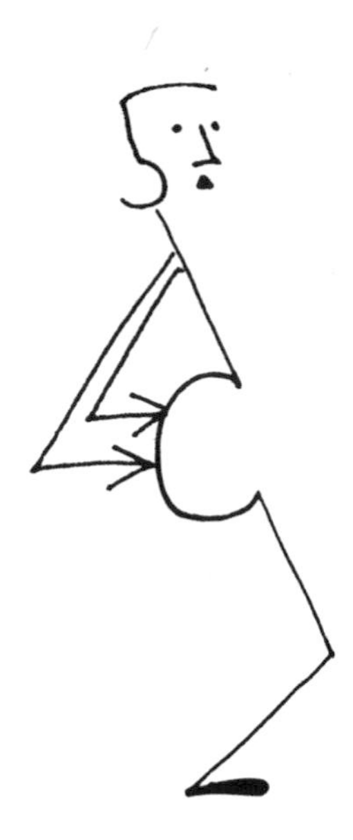

Keep pushing in with your fists. It probably hurts, but don't straighten up yet, and don't rock back and forth. That pain is an expression of your resistance to a lot of sensation. Don't try to loosen the resistance; just experience it. You may also feel some vibrations and, if you do, see what you learn.

The next step is to breathe deep down, below the belly button. Pull the air down and in. Do this for perhaps two minutes. Don't be in a hurry.

After several minutes, let out whatever sound wants to come out. I find "Aaah!" or "Oooh!" effective, but it may be a growl, a cry, a scream. See if you can experience it swelling up from your belly and pelvis.

Say your name. Keep saying it for a while and accept what you feel. It may be shakiness. It may be pleasure. It may be distrust. These are expressions of how you are grounded at this moment. Do you have feelings in your genitals? In your buttocks? In your feet?

Now straighten up without locking the knees. Put your hands on your head. Just stand and let yourself settle into your experiencing.

When you sit down, please keep both feet on the ground.

• • •

Let's continue.

Stand with your feet not too far apart, toes slightly pointed in. Bend all the way over, knees loose, head down, and put your fingertips to the floor. Don't lean forward; keep your weight back on your legs.

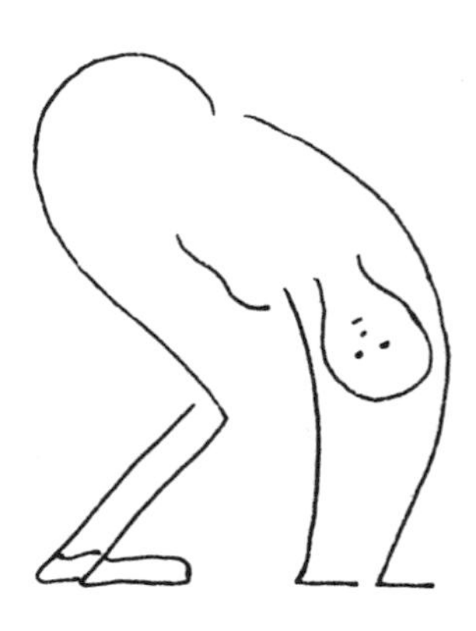

Now, by straightening the knees a little bit and loosening them again, raise yourself up and down a quarter of an inch or so. Let the belly and guts hang. And now begin to breathe the same as you did in the first exercise: deep into the abdomen. Feel your breathing expand into the lower back and buttocks. Feel your buttocks and thighs and calves and feet begin to pulsate with the in and out of the breathing. This pulsating opens up communication and connectedness with the ground.

If pain or discomfort begins to be noticeable, recognize it as a non-verbal expression of *no*. It expresses your resistance, your *no* to being connected to yourself and to the ground. At this point begin to verbalize the *no*. Make it conscious and take responsibility for it.

Keep breathing into your guts. Now straighten the knees another quarter of an inch into the pain. Make sure that you are still breathing into your belly. Let your buttocks flare open. Keep breathing. See if you can feel your legs and buttocks start to vibrate. If you feel more pain, say *no* again.

Now drop to a kneeling position and let your head rest on the floor. Let your whole backside open and close, expand and contract, as you breathe.

Very gradually find your way back to the original bent-over posture, with your feet and fingertips on the ground. And then find your way to a standing position.

Sit down when you want to. Remember to keep your feet on the floor.

• • •

The instinctive reaction to falling is to roll up into a ball, to assume the foetal position. Both of the above exercises are designed to make you aware that in order to block out sensations of falling and other dangerous feelings, you tighten up your stomach and legs. Without feelings in the front of your body and in the bottom half of your body, you don't have a good connection with the ground. You're confused about what you give and take in relation to the earth. Without a body that pulsates, you can easily believe that tight legs and a tight abdomen are normal. You identify with a feeling of holding back rather than with a feeling of down and out. Many people think that tightness represents strength. What it represents is the practicing of control, the imposing of the will. To relax threatens the stiff sense of self-mastery.

Most of us have fear in the vicinity of the anus and genitals. Our bottoms have been the recipients of both loving and disciplining. Generally speaking, in our culture the last vestige of intimate physical contact between parent and child occurs when mother powders the behind as she changes the diapers. Then comes toilet training and shame. We construct a social reward-system around the idea of control.

There is nothing wrong with learning control. There is nothing wrong with having the ability to control ourselves. But there's trouble when we can't stop controlling ourselves, when we can't give the bloody thing up. We can't make it back to the ocean of feeling; we are no longer participating with our feeling selves. And control is then the one thing that makes us feel potent. That part of us which controls us—which observes, analyzes, and criticizes—comes to have a stake in our not feeling grounded, in our not feeling connected.

Self-forming goes together with how we say *no.* Our *no* is dependent for its life upon the energy of the body.

If we connect with our sensations and feelings, we find that our controller is rooted in our pleasure, that it participates in our pleasure. Then it is no longer a controller at all. It becomes a tool for enhancing our enjoyment and self-development rather than for criticizing and restricting us.

Repeat these exercises, and see if you can gain more pleasure for yourself, more excitement to form yourself with.

From out of the great cornucopia of God's wanting, form finds itself.

CENTER FOR ENERGETIC STUDIES

The Center for Energetic Studies in Berkeley, California, under the direction of Stanley Keleman, is concerned with the study of the life of the body. The Center is interested in the biodrama of our lives, with how the rhythms and the cycles of feeling and need form our bodies and our lives.